W9-BAE-907

SOLID-STATE ELECTRONICS

FRANK P. TEDESCHI · MARGARET R. TABER

VNR VAN NOSTRAND REINHOLD COMPANY
NEW YORK CINCINNATI ATLANTA DALLAS SAN FRANCISCO

Van Nostrand Reinhold Company Regional Offices:
New York Cincinnati Atlanta Dallas San Francisco

Van Nostrand Reinhold Company International Offices:
London Toronto Melbourne

Library of Congress Catalog Card Number: 76-13561
ISBN 0-442-28460-8

Manufactured in the United States of America

Published by Van Nostrand Reinhold Company
450 West 33rd Street, New York, N.Y. 10001

Published simultaneously in Canada by Van Nostrand Reinhold Ltd.

15 14 13 12 11 10 9 8 7 6 5 4 3 2

Library of Congress Cataloging in Publication Data

Tedeschi, Frank P
Solid-state electronics.

(Electronics technology series)
Includes index.
1. Semiconductors. 2. Electronics
I. Taber, Margaret R., joint author. II. Title.
TK7871.85.T35 621.3815'3'042 76-13561
ISBN 0-442-28460-8

DEDICATION

I dedicate this book to my wife Marie whose understanding and patience have assisted me in writing this text. Also, I must mention my children Peter, John, Mary, and Catherine who had to survive many a weekend without a father, while he was completing this work.

FRANK P. TEDESCHI

I dedicate this book to my husband William, and to my students.

MARGARET R. TABER

Electronics Technology Series
Consulting Editor
Richard L. Castellucis

Foreword

In recent years, there has been a substantial increase in the demand for skilled electronics technicians due to the continued growth of the electronics industry, and the resulting increase in the development of electronic devices, systems, and appliances for industry, research, medicine, and the consumer. The need for technicians to design, build, test, install, service, repair, and replace these electronic components, devices, and systems is being met by numerous electronics technology instructional programs. These programs are found in several different settings including the technical high school, adult continuing education, industrial inservice training, and the two-year community (junior) colleges and technical institutes.

The authors and editors of this series established the following primary goals: readability, accuracy, currency, and reader involvement. Each text represents the current state of the art in electronics technology as it is commonly practiced in industry. Electronics theory is presented in each text to support the concepts which are to be applied to practical situations. The required level of mathematics does not go beyond algebra and trigonometry. Finally, each text is designed to involve the reader in a total learning experience. Each text unit lists the competencies that the reader is expected to demonstrate after mastering the concepts given in the unit. Extended Study Topics are presented to challenge motivated readers to go on and pursue more complex areas of interest. In addition, each text includes Laboratory Exercises to provide you with experience in working with actual electronic components in practical circuits.

To verify the technical content, each title in this series was classroom tested over a period of several years with two-year electronics technology students.

This Electronics Technology Series consists of the following titles:

Circuit Concepts: Direct and Alternating Current
Thomas S. Kubala

Solid-State Electronics
Frank P. Tedeschi and Margaret R. Taber

AC Circuit Analysis
Noble L. Lockhart and Ora E. Rice

Pulse and Logic Circuits
Richard L. Castellucis

Preface

SOLID-STATE ELECTRONICS is designed for use in junior colleges, technical institutes, and industrial training courses preparing technicians for careers in electronics. The material in the text furnishes the reader with basic semiconductor and electronic principles. These principles must be mastered before progressing to sophisticated semiconductor principles such as feedback amplifiers, field-effect transistors, integrated circuits, and SCRs. The only prerequisite to the use of this text is a sound knowledge of algebra and basic electricity.

The text is divided into eleven units. Each unit presents theory and includes numerous example problems followed by problems to be worked by the reader. Included at the end of each unit are laboratory experiments that are used as practical reinforcement of the theoretical ideas presented earlier in the unit. The experiments are designed so that they can be performed in most electrical laboratories. The experiments benefit the reader in that confidence is established in working with actual semiconductor components. The experiments are designed to provide an understanding of the basic concepts of circuit operation; hence, they are relatively simple.

Every effort has been made to adhere to the IEE Standard No. 260. Symbols for Units, adopted in 1965. The authors strongly encourage the use of the symbols presented in the text for their course work.

M.R. Taber and F. Tedeschi

Cleveland, Ohio

Contents

Unit 1 Introduction to *p-n* junctions and the semiconductor diode . . . 1

Electron forces and energies
Molecular bonding
Energy bands in solids
n-type semiconductor
p-type semiconductor
p-n junction
Semiconductor diode
Laboratory Exercise 1-1: Analysis of semiconductor diodes

Unit 2 Rectification 21

Diode used as a rectifier
Ideal transformer theory
Two-diode full-wave rectifier
Full-wave bridge rectifier circuit
Laboratory Exercise 2-1: Half-wave rectifier circuit
Laboratory Exercise 2-2: Transformer
Laboratory Exercise 2-3: Full-wave two-diode rectifier circuit
Laboratory Exercise 2-4: Full-wave bridge rectifier circuit

Unit 3 Power supply filters 41

The RC filter
Ripple voltage, ripple factor, and voltage regulation
RC filter design problems
Voltage doubler circuit
Choke input or L section filter
Critical inductance for L filters
Laboratory Exercise 3-1: RC power supply filter
Laboratory Exercise 3-2: Voltage doubler

Unit 4 Zener diodes 62

Zener diode theory
Zener diode ratings and specifications
Zener diode voltage regulator circuits
Laboratory Exercise 4-1: Zener diode voltage regulator

Unit 5 Junction transistor familiarization 77

Junction transistor action
Common-base characteristic curves
Laboratory Exercise 5-1: Transistor familiarization
Laboratory Exercise 5-2: CB collector characteristics

Unit 6 Common-base configuration 93
Simple common-base amplifiers
Common-base input and output resistance and voltage gain
Laboratory Exercise 6-1: Common-base amplifier
Unit 7 Common-emitter and common-collector amplifiers 105
Common-emitter characteristic curves
Simple common-emitter amplifiers
Simple common-collector amplifiers
Laboratory Exercise 7-1: CE collector characteristics
Laboratory Exercise 7-2: Common-emitter amplifier
Laboratory Exercise 7-3: Common-collector (emitter-follower) amplifier
Unit 8 Transistor specifications and graphical analysis 132
Transistor specification sheets and terminology
Power dissipation curve
Unit 9 Transistor load lines 149
Introduction
Dc load line
Ac load line
Laboratory Exercise 9-1: Load line experiment
Unit 10 Transistor biasing for common-emitter circuits 168
Introduction
Fixed-bias circuit
Emitter-bias with single base resistor
Laboratory Exercise 10-1: Fixed-bias circuit
Laboratory Exercise 10-2: Emitter-bias circuit
Unit 11 More common-emitter biasing 181
Emitter-bias circuit with voltage divider
Collector-base bias
Collector-base bias circuit with emitter resistor
Emitter-bias circuit with two supplies
Laboratory Exercise 11-1: Emitter-bias circuit with voltage divider
Laboratory Exercise 11-2: Collector-base bias circuit
Laboratory Exercise 11-3: Emitter-bias circuit with two supplies

Appendix: Answers to Student Review Questions 198

Acknowledgment Page 202

Index 203

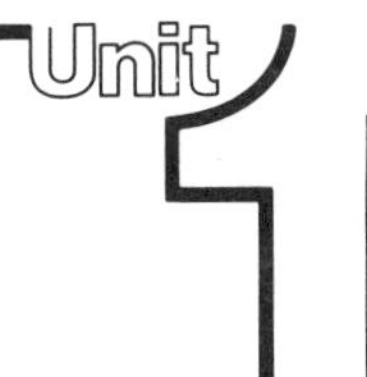

Introduction to p-n junctions and the semiconductor diode

OBJECTIVES

After studying this unit, the student will be able to discuss and demonstrate an understanding of the basic principles of:

- The solid-state physics governing the phenomenon occurring in semiconductors
- Problem solving related to the solid-state physics of semiconductors
- Semiconductor diode characteristics

ELECTRON FORCES AND ENERGIES

The electrons of an atom move in dynamically stable orbits about the nucleus of the atom. Since the electrons have a mass and are moving about the nucleus, they possess a centrifugal force. In addition, there is a centripetal electrostatic force between the electrons in their orbits and the protons in the nucleus. This electrostatic force is described by Coulomb's law for charged bodies.

Figure 1-1, page 2, shows a hydrogen atom consisting of one electron and one proton. The forces acting on the electron and the velocity of the electron are indicated. The electrostatic force F_e exerted by the proton on the electron provides the centripetal force required to hold the electron in a circular orbit about the nucleus. Since the proton is nearly two thousand times heavier than the electron, the motion of the proton is not under the influence of the electrostatic force exerted by the electron. The electron's electrostatic force can be compared to the earth's effect upon the sun's motion.

The centrifugal force exerted on the electron is expressed by the equation

$$F_c = \frac{mv^2}{r} \qquad \text{Eq. 1.1}$$

where r = radius from the proton to the orbit of the electron, in meters

v = velocity of the electron, in meters per second

m = mass of the electron, in kilograms

F_c = centrifugal force, in newtons

Coulomb's law for charged bodies is used to determine the electrostatic force.

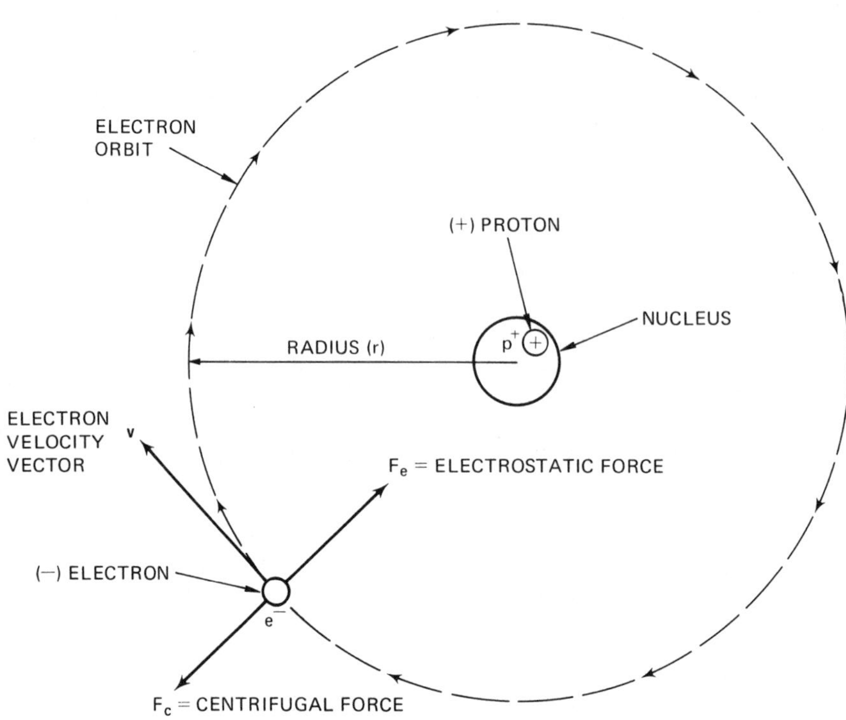

FIG. 1-1 THE FORCES ACTING ON THE ELECTRON AND THE VELOCITY OF THE ELECTRON IN THE HYDROGEN ATOM

$$F_e = 9 \times 10^9 \frac{q_1 \, q_2}{r^2} \qquad \text{Eq. 1.2}$$

where q_1 = charge on body one, in coulombs

q_2 = charge on body two, in coulombs

r = distance between q_1 and q_2, in meters

F_e = force, in newtons

The charges of the electron and proton are equal in value but opposite in charge. This means $q_1 = q_2 = e = 1.6 \times 10^{-19}$ coulombs (e is the charge of the electron). As a result, the electrostatic force exerted on the electron by the proton is

$$F_e = 9 \times 10^9 \frac{e^2}{r^2} \qquad \text{Eq. 1.3}$$

Since the electron remains in its orbit about the nucleus, the electrostatic force exerted by the proton on the electron must equal the centrifugal force of the electron. Therefore, Eq. 1.3 and Eq. 1.1 can be equated:

$$\frac{mv^2}{r} = 9 \times 10^9 \frac{e^2}{r^2}$$

By rearranging the terms in this relationship, it is possible to determine the velocity of the electron in relation to the radius of its orbit.

$$v = \frac{e}{\sqrt{\dfrac{mr}{9 \times 10^9}}} \qquad \text{Eq. 1.4}$$

The kinetic energy (KE) of the electron can be found using the equation

$$KE = \frac{1}{2} m v^2 \qquad \text{Eq. 1.5}$$

where m = mass of the electron, 9.11×10^{-31} kilograms

v = velocity of the electron, in meters per second

KE = kinetic energy of the electron, in joules

The electrostatic potential energy (PE) of the electron is determined by multiplying the electrostatic force (Eq. 1.2), by the orbit radius and by the quantity minus one:

$$PE = -F_e r$$

or **Eq. 1.6**

$$PE = -9 \times 10^9 \frac{e^2}{r}$$

where e = charge of an electron, 1.6×10^{-19} coulombs

r = radius from proton to electron (orbit radius), in meters

PE = electrostatic potential energy, in joules

The electrostatic potential energy is negative due to the fact that the electron is attracted to the much heavier proton. This means that the potential energy found by Eq. 1.6 cannot be used by the electron to do work on anything outside the atom.

The total electron energy E for the hydrogen atom is equal to KE plus PE.

$$E = KE + PE = \frac{mv^2}{2} - 9 \times 10^9 \frac{e^2}{r}$$

Substituting the electron velocity v from Eq. 1.4 into this equation yields:

$$E = \frac{m}{2}\left(\frac{e^2}{\frac{mr}{9 \times 10^9}}\right) - 9 \times 10^9 \frac{e^2}{r}$$

$$E = 9 \times 10^9 \left(\frac{e^2}{2r} - \frac{e^2}{r}\right)$$

$$E = -4.5 \times 10^9 \frac{e^2}{r}$$

The total energy of the electron is negative, indicating that the electron is bound to the nucleus. A positive E means that the electron is not able to remain in a closed orbit about the nucleus.

PROBLEM 1

Experiments indicate that 2.2×10^{-18} joules are required to separate a hydrogen atom into a proton and an electron. In other words, the binding energy of a hydrogen atom is 2.2×10^{-18} joules. Find the values for the orbital radius and the velocity of the electron using the given binding energy.

$$e = 1.6 \times 10^{-19} \text{ coulombs}$$

$$m = 9.11 \times 10^{-31} \text{ kilograms}$$

Use Eq. 1.7 to find the radius of the electron orbit:

$$r = \frac{-4.5 \times 10^9 \times (1.6 \times 10^{-19})^2}{-2.2 \times 10^{-18}}$$

$$= 5.3 \times 10^{-11} \text{ m}$$

Note that the binding energy E is negative in the above equation, since the electron is bound to the nucleus. The electron velocity v is found using Eq. 1.4:

$$v = \frac{e}{\sqrt{\frac{mr}{9 \times 10^9}}}$$

$$= \frac{1.6 \times 10^{-19}}{\sqrt{\frac{9.11 \times 10^{-31} \times 5.3 \times 10^{-11}}{9 \times 10^9}}}$$

$$v = 2.19 \times 10^6 \text{ meters per second (m/s)}$$

PROBLEM 2

The electron of the hydrogen atom can exist in various orbits about the hydrogen nucleus. Plot a graph of KE, PE, and E for differing values of the orbital radius for the hydrogen atom.

To obtain the values for these quantities, substitute the values for r in Eq. 1.5, 1.6, and 1.7. Figure 1-2 illustrates the resulting graphs. Figure 1-2 shows that the kinetic energy of the electron *decreases* as it moves away from the nucleus. The electrostatic potential energy *increases* by twice this value, resulting in a net increase in total energy. Remember that negative values of small magnitudes are greater than negative values of larger magnitudes.

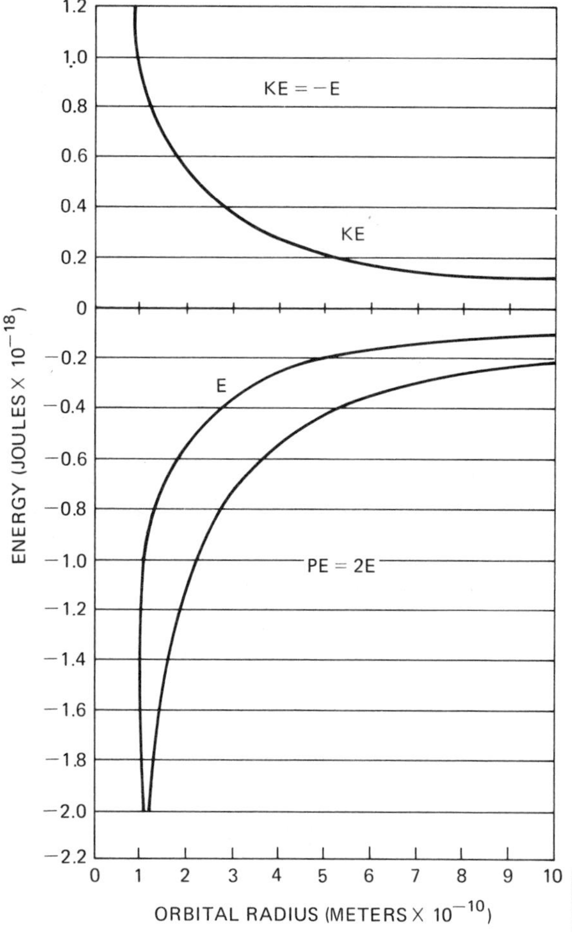

FIG. 1-2 ENERGY PROFILES IN THE HYDROGEN ATOM

Calculate the kinetic energy, potential energy, and total energy of an electron whose orbit is 9 x 10^{-10} meters. (R1-1)

Find the velocity of the electron in question (R1-1). (R1-2)

MOLECULAR BONDING

When atoms are joined together they form molecules of matter. The joining process is called *bonding,* and occurs through the interaction of the *valence* electrons of each atom. The forces which bind the atoms are electrostatic in nature and are divided into three main categories: (a) *ionic* bonding forces, (b) *covalent* bonding forces, and (c) *metallic* bonding forces. Ionic bonding occurs when valence electrons are *transferred* from one atom to another. Covalent bonding occurs when one or more pairs of valence electrons are *shared* by atoms. Metallic bonding occurs when valence electrons *float in a cloud* among positive ions of metallic atoms.

Covalent bonding between atoms results from the sharing of the valence electrons of the atoms. Because of the sharing process, the atoms do not become ionized but remain neutral.

Figure 1-3 illustrates valence electron sharing for the hydrogen molecule H_2. Each electron spends an equal amount of time or-

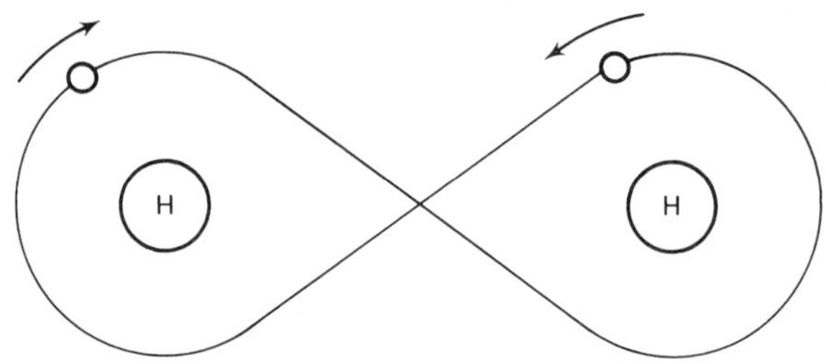

FIG. 1-3 COVALENT BONDING OF TWO HYDROGEN ATOMS

biting each nucleus. The force of attraction exerted by both nuclei on the electron is the attractive force holding the hydrogen atoms together. This attractive force is counter-balanced by direct repulsion between the nuclei of the two atoms. If the nuclei are too close together, their mutual repulsion dominates and the hydrogen molecule is not stable.

Germanium and silicon are two elements whose molecules are held together by covalent bonding. Both elements possess four valence electrons. In the covalent bonding of molecules of these elements, each atom shares one of its valence electrons with each of four neighboring atoms.

The covalent bonding of pure (*intrinisc*) germanium is shown in figure 1-4. The outer shell for germanium is the n-shell which has four valence electrons. A germanium molecule is stable when each germanium atom shares its four valence electrons with four neighboring germanium atoms. At the same time, each of the neighboring atoms shares one of its valence electrons with the central atom. This type of electron sharing is called *electron-pair bonds.* In other words, covalent forces are established when two electrons coordinate their activities so as to produce an electrostatic force between the electrons. Figure 1-5 illustrates a two-dimensional representation of covalent bonding by germanium atoms.

The student must understand the principle of covalent bonding because of the importance of this concept in describing the physical action of semiconductors (covered later in this unit).

Define the following quantities. *(R1-3)*

a. ionic bonding

b. covalent bonding

c. metallic bonding

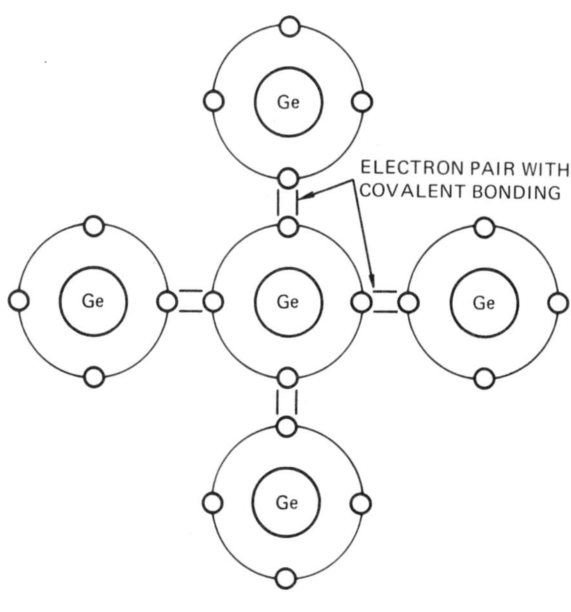

FIG. 1-4 COVALENT BONDING OF A GERMANIUM ATOM WITH FOUR NEIGHBORING ATOMS

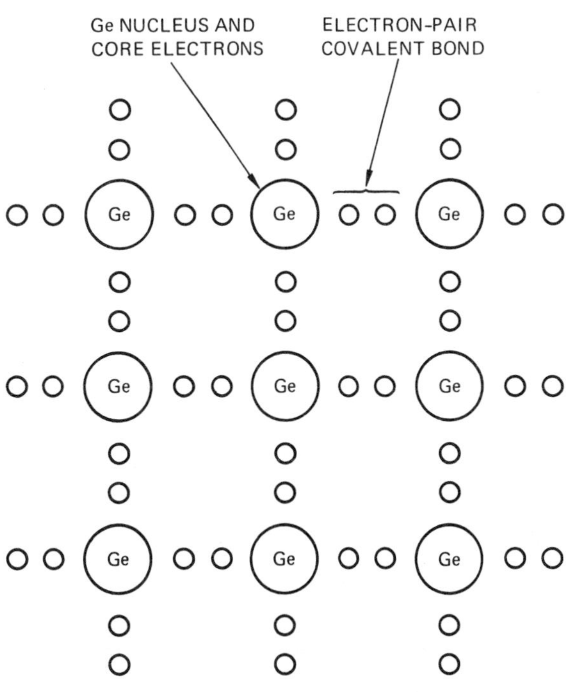

FIG. 1-5 THE ELECTRON-PAIR COVALENT BONDING OF THE MOLECULAR STRUCTURE OF GERMANIUM (Ge)

ENERGY BANDS IN SOLIDS

The energy equations derived earlier in this unit (equations 1.5, 1.6, and 1.7), are based upon the *particle* properties of the electron. The particle nature of matter obeys the ordinary (Newton's) laws of physics. However, there are some electron phenomena which cannot be explained by the ordinary laws of physics. The explanation of these phenomena is given by a branch of physics called *quantum mechanics.* We now know that electrons have a *dual* nature; that is, they have both particle-like properties and wavelike properties. According to quantum mechanics, the energies associated with an electron depend upon an energy shell *n*. The quantity n is called the *quantum number.* The following equation expresses electron energy in terms of the quantum number.

$$E_n = -5.09 \times 10^{20} \frac{m_e \; e^4}{n^2 \; h^2} \qquad \text{Eq. 1.8}$$

where m_e = mass of an electron, in kilograms

e = charge of an electron, in coulombs

h = Planck's constant, 6.63×10^{-34} joule-second

n = quantum number, 1, 2, 3, ...

E_n = electron energy, in joules

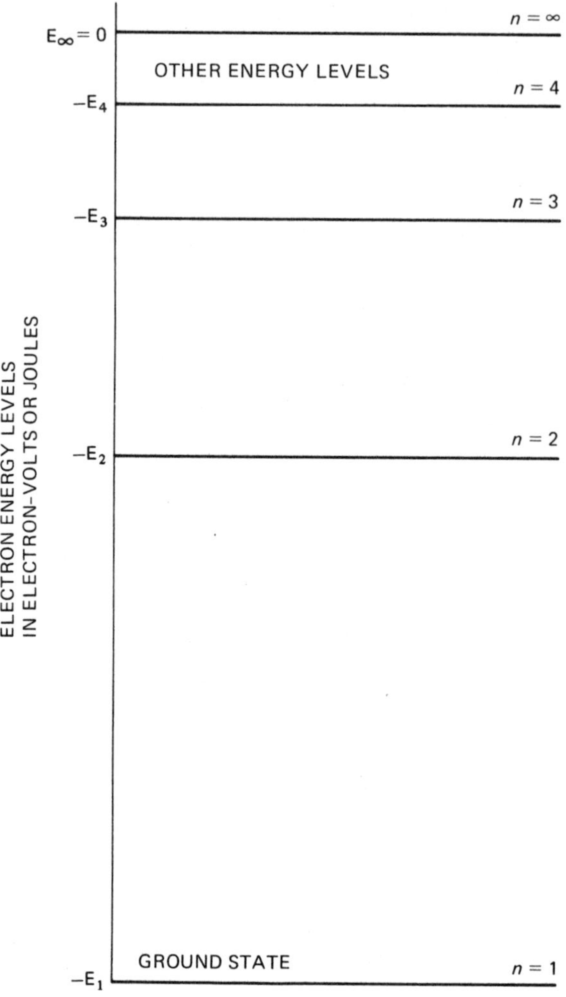

FIG. 1-6 TOTAL ENERGY LEVEL DIAGRAM FOR ELECTRONS IN TERMS OF QUANTUM NUMBERS *n*.

The values of electron energies obtained by the use of Eq. 1.8 with different values of n are shown in figure 1-6. This diagram is called an *energy level* diagram. Note that all of the energies are negative. The lowest energy level E_1, corresponding to quantum number n = 1, is called the *ground state* of the atom. The ground state condition exists only when the temperature of the immediate environment around the atom is equal to absolute zero. At all other temperatures, the electrons are agitated and moving continuously between allowable energy levels. In such a state, an atom is said to be *excited.* When moving between energy levels, an electron experiences step-type changes (discrete changes) in its energy due to the absorption or emission of units of thermal energy called *photons* or *quanta.*

When many atoms are brought close together as in a solid, the energy levels of the individual atoms form *bands of energies.* Numerous energy bands are formed, but the energy bands that determine the electrical properties of matter are the *valence band, conduction band,* and *forbidden band* (or gap), figure 1-7.

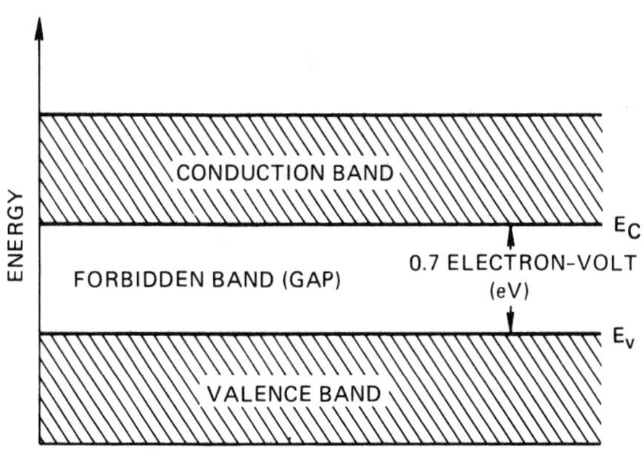

FIG. 1-7 ENERGY BAND DIAGRAM OF GERMANIUM

The valence energy band contains the energy levels of all of the valence electrons in a crystal structure. The electrons in the valence band are attached to parent atoms and are not free to move about unless supplied with energy from an external source. The valence band electrons can be elevated into the conduction band if the proper amount of external energy is added.

To move into the conduction band, the valence electrons must bridge an energy gap called the *forbidden* energy band. No electrons can occupy the forbidden band which determines whether a solid will act as a conductor, a semiconductor, or an insulator.

In the conduction energy band, the energy of the electrons is high enough so that electrons in this band are not attached or bound to an atom but are *free* or *mobile.* Electrons normally do not exist in the conduction band. However, electrons do move to this band from the valence band after they gain sufficient energy to cross the forbidden gap. Electrons that reach the conduction band are capable of being influenced by an external force.

A conductor is a solid in which the valence band and the conduction band overlap at room temperature. This means that valence electrons can be found in the conduction band at room temperature. There is no forbidden gap between the valence and conduction bands as shown by the conductor energy diagram in figure 1-8a. As a result, conductors have free electrons in the conductor band. These electrons are available to conduct current without first requiring that external energy be applied to the valence electrons to move them into the conduction band. The energy diagram of an insulator is shown in figure 1-8b. Note that the forbidden gap is very wide. Under normal conditions, this means that few

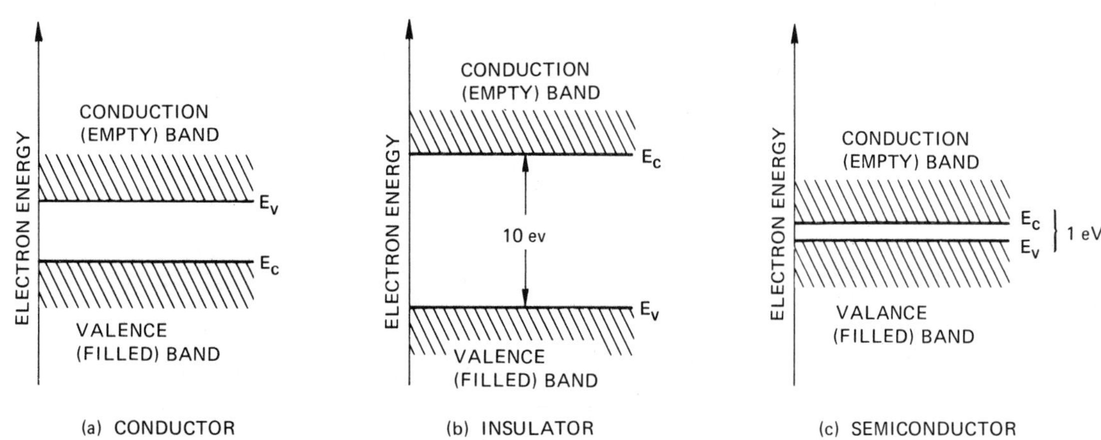

FIG. 1-8 ENERGY DIAGRAMS FOR A CONDUCTOR, AN INSULATOR, AND A SEMICONDUCTOR

valence electrons can be given sufficient energy to bridge the gap between the valence band and the conduction (free) band. Ideally, then, an insulator has an unoccupied conduction band and a completely occupied valence band. However, since there are no perfect insulators, a few electrons will exist in the conduction band.

A semiconductor is a solid which has a forbidden energy band that is smaller than that of an insulator and larger than that of a conductor. An average value for the energy of the forbidden gap of a semiconductor is 1 electron-volt (eV) as shown in figure 1-8c. At room temperature, enough energy is provided to the valence electrons of semiconductors to enable them to cross the forbidden gap and exist as conduction (free) electrons in the conduction band. Thus, semiconductors can conduct some electric current at room temperature. Semiconductors are the solids used in solid-state devices such as transistors, tunnel diodes, and field-effect transistors.

When working with semiconductors or other solids, there are two types of current carriers to be considered, (a) *electrons* and (b) *holes*. For now, a hole is defined as an electron *vacancy*. In other words, if a valence electron moves from the valence band to the conduction band, then the electron vacancy in the valence band is called a hole. Thus, as valence electrons move in a solid, there is also apparent hole motion. In many respects, holes act as positive charges since they attract negative charges such as electrons.

A good conductor is characterized electrically by its low value of resistivity (ρ) and its high value of conductance (σ). The equation for conductance is:

$$\sigma = q(n\mu_n + p\mu_p) \qquad \text{Eq. 1.9}$$

where σ = the conductance of a solid, in mhos per centimeter

q = the charge of an electron

n = the number of electrons available for conduction, in electrons per cubic centimeter

p = the number of holes available for conduction, in holes per cubic centimeter

μ_n = the electron mobility factor (given by equation 1.10)

μ_p = the hole mobility factor (given by equation 1.10)

The mobility factor, μ, indicates the relative ease with which an electron or hole can move through the structure of a solid. The mobility factor is a function of the arrangement of the atoms in the crystal structure of the solid, the dielectric constant, the amount of vibration shown by the atoms, and the temperature. For a specific material, the mobility factor is found with the use of equation 1.10.

$$\mu = BT^{-1.5} \qquad \text{Eq. 1.10}$$

where B = constant for a given solid

T = absolute temperature, in degrees Kelvin (°K)

μ = electron or hole mobility factor; this factor is a constant for a given temperature, in square centimeters per volt-second

Table 1-1 lists some of the properties of intrinsic germanium and silicon.

PROBLEM 3

Intrinsic germanium at a temperature of 300°K has 4.1×10^{22} electrons/cm^3 available for conduction. Find the conductivity and resistivity of the germanium structure.

The mobility factors for germanium are given in Table 1-1:

μ_n = 3900 and μ_p = 1900.

n = p = 4.1×10^{22} for intrinsic germanium.

$\sigma = q(n\mu_n + p\mu_p$

$\sigma = 1.6 \times 10^{-19} (4.1 \times 10^{22} \times 3900 + 4.1 \times 10^{22} \times 1900)$

$= \mathbf{3.81 \times 10^7}$ **mhos/cm**

Since resistivity is the reciprocal of conductivity,

$\rho = \frac{1}{\sigma} = \frac{1}{3.81 \times 10^7}$

$\mathbf{2.625 \times 10^{-8}}$ **= ohm-cm**

A semiconductor has a higher value of resistivity than a conductor. For example, a cubic centimeter of silicon has about one current carrier for each four trillion (4×10^{12}) carriers in a cubic centimeter of copper. This tremendous difference in current carriers is explainable in terms of the widths of the energy bands for a semiconductor and a conductor.

For normal atomic spacing, the filled covalent levels of silicon are separated from the nearest empty levels by a forbidden gap of approximately 1.11 eV. The value of the forbidden gap for germanium is 0.7 eV. In other words, a covalent electron in silicon must receive energy equal to at least 1.11 eV if it is to be moved into the conduction band and become a free electron; covalent electrons of germanium require at least 0.7 eV of energy to jump across the forbidden gap to become conduction (free) electrons. There is usually enough thermal energy at room temperature supplied to covalent silicon or germanium electrons to enable them to cross the forbidden gap into the conduction band.

When an electron does jump from the valence band into the conduction band, a vacancy is left in the valence band. This vacancy is called a hole and has a strong affinity for an electron because of the attraction between opposite charges. A hole can move from one atom to another in the valence band by capturing a covalent electron from an adjacent atom. The atom that loses the electron gains the hole. In this manner, a hole moves in the valence band along a random path from atom to atom, figure 1-9, page 10.

For an intrinsic (pure) solid material, the number of electrons in the conduction band is equal to the number of holes in the valence band. Because of the fact that electrons and

TABLE 1-1 PROPERTIES OF INTRINSIC GERMANIUM AND SILICON

	Germanium	Silicon
Atomic Number	32	14
Valence Electrons	4	4
Atoms/cm^3	4.4×10^{22}	5×10^{22}
Energy gap at 81° F, in eV, E_g	0.72	1.1
Energy gap at -460° F, in eV, E_g	0.785	1.21
Number of charge carriers in intrinsic material, n_i; carrier density at 300°K for electrons (n)/cm^3 or holes (p)/cm^3	2.5×10^{13}	1.5×10^{10}
Electron mobility, μ_n, at 300° K or 81° F cm^2/volt-second	3900	1500
Hole mobility, μ_p, at 300° K or 81° F cm^2/volt-second	1900	500
Diffusion constant for electrons, D_n, at 81° F cm^2/second	99	39
Diffusion constant for holes, D_p, 81° F cm^2/second	49	12
Intrinsic resistivity at 81° F, ohm-cm	47	2.1×10^5
Boltzmann's constant (k) = 1.38×10^{-23} joule/K°		
Planck's constant (h) = 6.624×10^{-34} joule-sec		
Charge of one electron (q) = 1.602×10^{-19} coulombs		

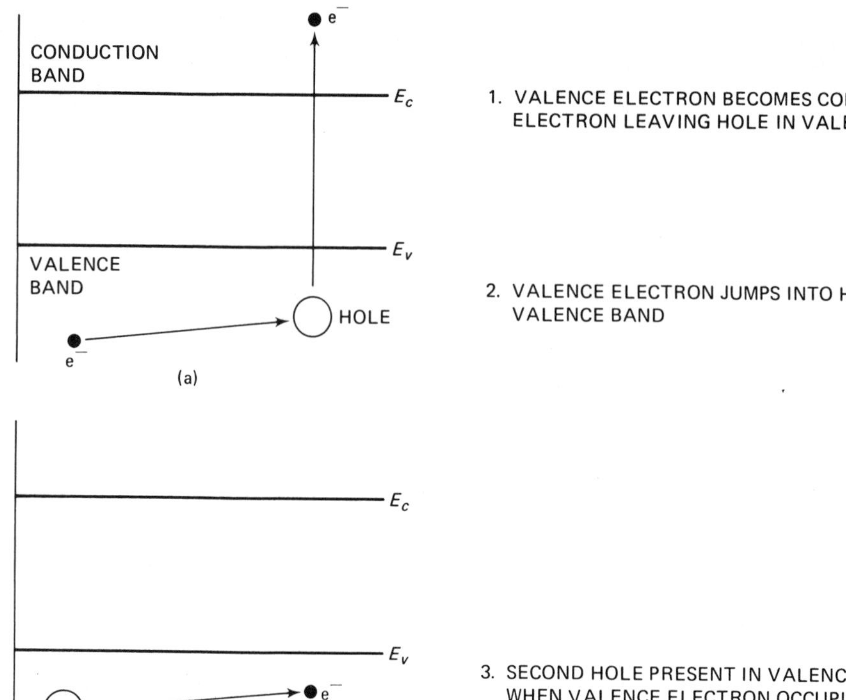

FIG. 1-9 HOLE AND ELECTRON MOTION IN THE VALENCE BAND

holes always generate and recombine in pairs, the phrase *electron-hole pairs* is used. The average number of electrons in the conduction band or the average number of holes in the valence band of an intrinsic material can be found by the following equation.

$$n_i = AT^{1.5} \epsilon^{-E_g \frac{q}{2KT}} \qquad \text{Eq. 1.11}$$

where n_i = the average number of electrons per cubic centimeter in the conduction band for intrinsic material

T = the absolute temperature, in degrees Kelvin

k = Boltzmann's constant, 1.38×10^{-23}

E_g = the width of the forbidden gap, in electron-volts (eV)

q = the charge of an electron

A = a constant for any given intrinsic material

The difference between semiconductors and insulators is in the width of the forbidden gap. The width of the gap for insulators may be several electron volts, whereas the width of the gap for semiconductors is approximately one electron volt. If the forbidden gap is greater than a few electron volts, there are very few free electrons in the conduction band. This is due to the fact that the thermal energy at room temperature is not sufficient to move valence electrons to the conduction band.

PROBLEM 4

An intrinsic material has a forbidden gap of 2.58 eV. Find the number of current carriers at room temperature (300° K) for this material if the constant A is equal to 10×10^{16}.

$$\frac{q}{kT} = \frac{1.6 \times 10^{-19}}{1.38 \times 10^{-23} \times 300} = \frac{1}{0.0258}$$

$$n_i = AT^{1.5} \ \epsilon^{-E_g/2 \times 0.0258}$$

$$n_i = 10 \times 10^{16} \times 300^{1.5} \times \epsilon^{-2.58/0.0516}$$

$$n_i = 5.18 \times 10^{20} \times \epsilon^{-50} = 0.1 \text{ electrons/cm}^3$$

Is silicon or germanium more sensitive to heat? Explain your answer. (R1-4)

Determine the conductivity and resistivity of intrinsic germanium at room temperature, if there are 2.5 x 10^{13} electrons per cubic centimeter available for conduction. (Refer to Table 1-1.) (R1-5)

Determine the conductivity and resistivity of intrinsic silicon, if there are 1.6 x 10^{10} electrons per cubic centimeter available for conduction. (Refer to Table 1-1.) (R1-6)

n-TYPE SEMICONDUCTOR

Group III elements in the periodic table contain three valence electrons; these elements are called *trivalent* elements. Group V elements in the periodic table contain five valence electrons; these elements are called *pentavalent* elements. If germanium or silicon is doped by adding pentavalent atoms, the two-dimensional crystal structure shown in figure 1-10 is obtained. To dope germanium or silicon means the addition of one impurity atom to 10^8 Ge or Si atoms. In figure 1-10, four of the five valence electrons of the pentavalent atom exhibit covalent bonding with neighboring germanium or silicon valence electrons. The fifth valence electron of the pentavalent atom is a *free electron* because it is not in the valence band and is not a part of the covalent sharing process.

A comparison of intrinsic and doped (*impure* or *extrinsic*) germanium or silicon at absolute zero is shown in figure 1-11. Note how close the free electrons are to the conduction band for doped germanium and sili-

FREE ELECTRON FROM PENTAVALENT IMPURITY ATOM

FIG. 1-10 *n*-TYPE SEMICONDUCTOR

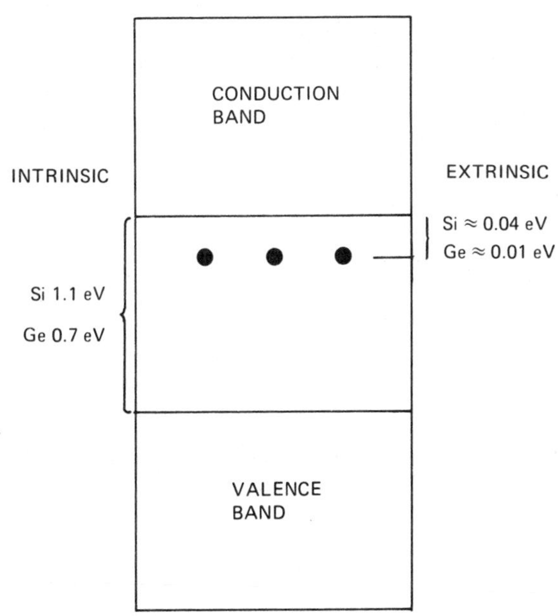

FIG. 1-11 COMPARISON OF INTRINSIC AND *n*-TYPE SEMICONDUCTORS AT ABSOLUTE ZERO

con. At room temperature, the free electrons formed by the doping process together with the electron-hole pairs of the basic pure semiconductor material actually are located in the conduction band ready to conduct when a potential is applied to the structure. Because extra or free electrons are added to the intrinsic structure, the resulting material is called an *n*-type semiconductor. The impurity used to dope the intrinsic germanium or silicon is called a donor because it donates an electron to the semiconductor structure. The basic *n*-type semiconductor is not electrically negative; it is neutral because the impurity atom added with five valence electrons also has five protons. At room temperature, the *majority carriers* in the *n*-type semiconductor are electrons. These electrons are supplied by the doping process and also are released from the valence band due to the action of the thermal energy. The electrons moving out of the valence band leave holes which are called *minority carriers.*

Define the following terms: (a) doping; (b) extrinsic germanium; (c) n-type majority carriers; (d) n-type minority carriers; (e) donor atoms. *(R1-7)*

Explain covalent bonding of Group V elements with intrinsic germanium or silicon. *(R1-8)*

p-TYPE SEMICONDUCTOR

If germanium or silicon is doped with trivalent atoms, the two-dimensional crystal structure shown in figure 1-12 is obtained. The three valence electrons from the trivalent doping atom form covalent bonds with four neighboring silicon or germanium atoms. However, one of the four covalent bonds lacks an electron. In other words, this semiconductor is said to be *doped with holes.*

Figure 1-13 compares intrinsic and extrinsic (doped) germanium and silicon at absolute zero. Note how close the holes are to the valence band for doped germanium and silicon. An electron in the valence band needs the addition of only a little energy to enable it to jump into the hole. With the electron in its new location, a hole is left in the valence

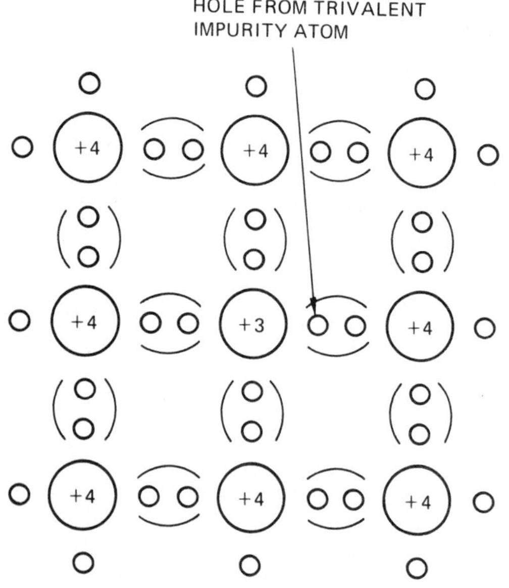

FIG. 1-12 *p*-TYPE SEMICONDUCTOR

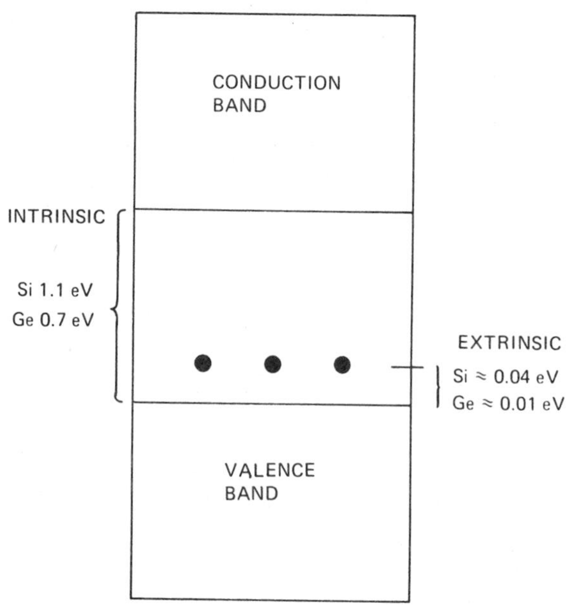

FIG. 1-13 COMPARISON OF INTRINSIC GERMANIUM AND SILICON WITH *p*-TYPE SEMICONDUCTORS

band. At room temperature, the doped holes actually are filled already, leaving holes in the valence band. The holes in the valence band and the few electron-hole pairs of the pure semiconductor material will flow when an electrical potential is applied. The addition of extra holes to the pure semiconductor results in a *p*-type semiconductor. The impurity added to the pure semiconductor material is called an acceptor-type impurity because it accepts electrons. The *p*-type semiconductor is not electrically positive; it is neutral due to the fact that the impurity atom added has three valence electrons and three protons. At room temperature, the majority carriers in *p*-type semiconductors are holes. The holes occur as a result of the process of doping and when some electrons in the valence band receive enough energy to move to the conduction band. A few electron-hole pairs generated by thermal energy do exist in a *p*-type semiconductor. Thus, a *p*-type semiconductor will have minority carrier electrons.

Define the following terms: (a) p-type majority carriers; (b) p-type minority carriers; (c) hole current; (d) acceptor atoms. (R1-9)

Explain the process of covalent bonding of Group III elements with intrinsic germanium or silicon. (R1-10)

p-n JUNCTION

A *p*-type semiconductor at room temperature is shown with a large number of holes and acceptor ions in figure 1-14. At room temperature, electrons fill the doped holes and leave holes in the valence band. The impurity atom has only three electrons, but it gains a fourth when the hole is filled. Since there are four electrons and only three protons, a negative ion is produced. Figure 1-14 also shows electron-hole pairs due to thermal agitation.

Figure 1-15 shows an *n*-type semiconductor at room temperature with a large number of electrons and donor ions. At room temperature, the doped electrons receive enough energy to exist in the conduction band. The doped electrons are provided by the impurity atoms with five valence electrons. When the fifth electron or free electron moves to the conduction band, it leaves behind a positive ion consisting of four electrons and five protons. It is for this reason that the number of positive donor ions equals the number of electrons in figure 1-15. Also shown in the figure are a few electron-hole pairs resulting from thermal agitation.

The two types of semiconductor materials can be joined as shown in figure 1-16, page 14, to form a *p-n junction diode.* This process is not a simple matter of combining *p* and *n* materials. There are a variety of complicated techniques for joining *p-n* junctions including grown junction, alloy junction, epitaxial layer, diffusion, and electrochemical etching and plating.

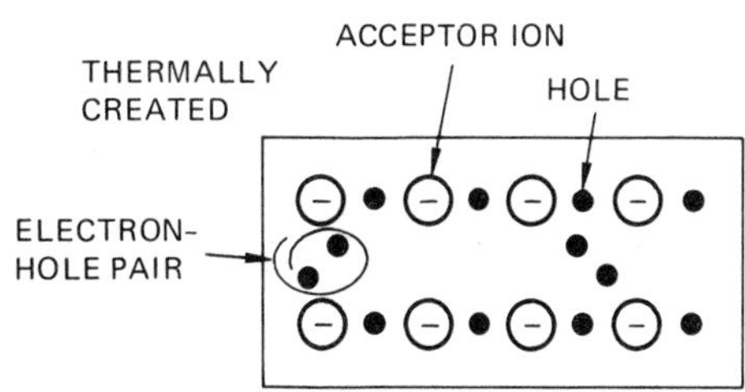

FIG. 1-14 *p*-TYPE SEMICONDUCTOR

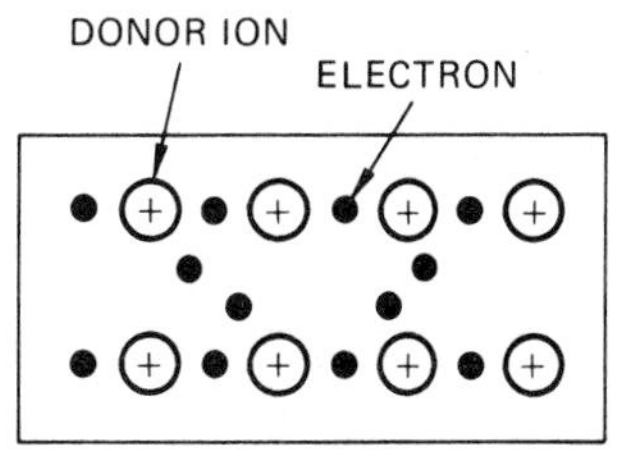

FIG. 1-15 *n*-TYPE SEMICONDUCTOR

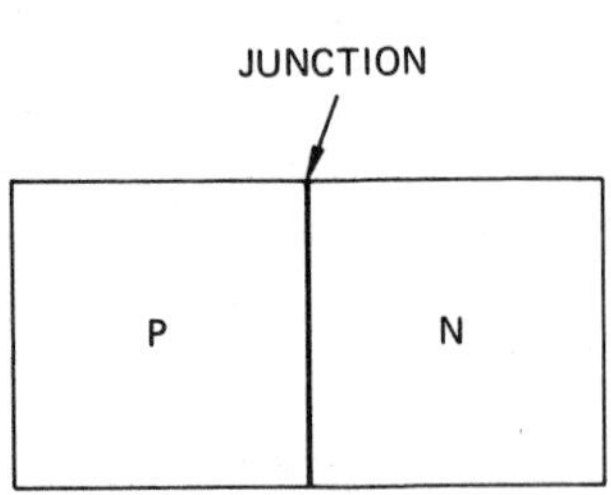

FIG. 1-16 *p-n* JUNCTION

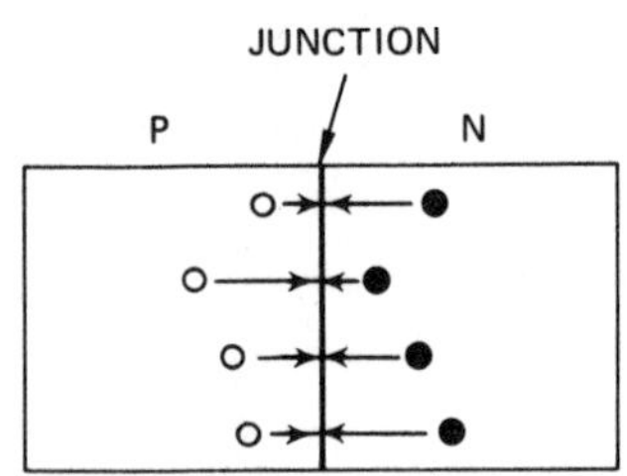

FIG. 1-17 ATTRACTION OF HOLES AND ELECTRONS TOWARD THE *p-n* JUNCTION

Figure 1-17 shows that when *p* and *n* materials are joined, the holes and electrons which carry opposite charges are attracted toward the *p-n* junction. Since ions are physically larger and heavier than the electrons and holes, they cannot move toward the junction. As a result, the ions maintain their positions in the crystal structure. The electrons and holes, however, move toward one another, cross the junction, and *recombine* to eliminate each other. After a few recombinations the process stops due to the presence of positive and negative ions near the junction. When an electron from the *n*-type material tries to cross the junction, it sees negative ions in the *p*-type material and is repelled by them. Similarly, holes from the *p*-type material are repelled by the positive ions in the *n*-type material.

A *barrier* now exists at the *p-n* junction since neither the electrons nor the holes can move across the junction. Note in figure 1-18 that the region around the junction lacks charge carriers. This region is called the *depletion region.* The physical distance of the depletion region is known as the *barrier width.* When an external potential is applied, there is a potential difference from one side of the depletion region to the other side. This potential difference is called the height of the barrier.

When an external battery is placed across a *p-n* junction, as in figure 1-19, the process is called biasing. The positive side of the battery in figure 1-19 is connected to the *n*-type material, and the negative side of the battery is connected to the *p*-type material. The holes in the *p*-type material see a negative potential; thus, they are attracted toward the negative potential and away from the junction. The electrons in the *n*-type material see a positive potential; they are attracted toward the positive potential and away from the junction. The movement of the electrons and holes away from the junction increases the barrier width. As a result, no current flows in the circuit due to the majority carriers. The minority carriers,

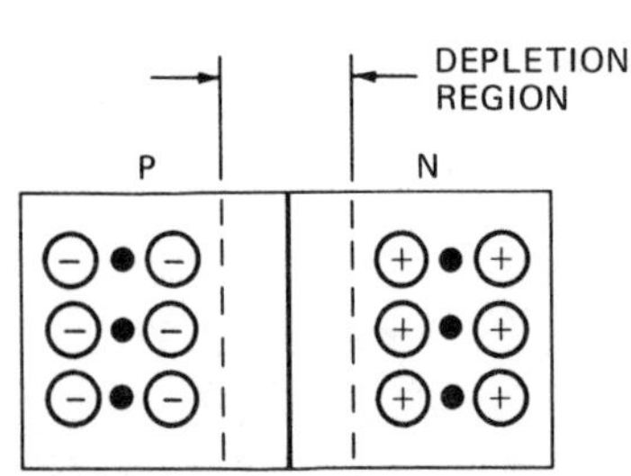

FIG. 1-18 DEPLETION REGION AROUND THE *p-n* JUNCTION

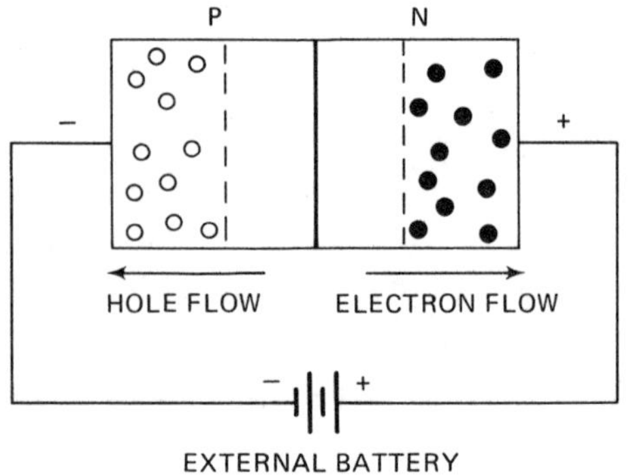

FIG. 1-19 REVERSE - BIASED *p-n* JUNCTION

however, are pushed toward and across the junction by the external potential of the battery. In other words, when a junction is *reverse biased* for majority carriers, it is *forward biased* for minority carriers. Thus, a very small current flows due to the minority carriers when a *p-n* junction is reverse biased as shown in figure 1-19. When a junction is said to be forward biased or reverse biased, it is understood that this means forward biasing or reverse biasing respectively for *majority* carriers.

Minority current flow is shown for a typical *p-n* junction diode on the *characteristic curve* in figure 1-20. The minority or *leakage current* is usually in the order of microamperes (μA) or nanoamperes (nA).

In figure 1-21, the positive side of the battery is connected to the *p* material and the negative side of the battery is connected to the *n* material. This is the forward bias condition. The majority carriers in the *p* material see a positive potential on the left and are repelled toward the junction. The electrons in the *n* material see a negative potential on the right and also are repelled toward the junction.

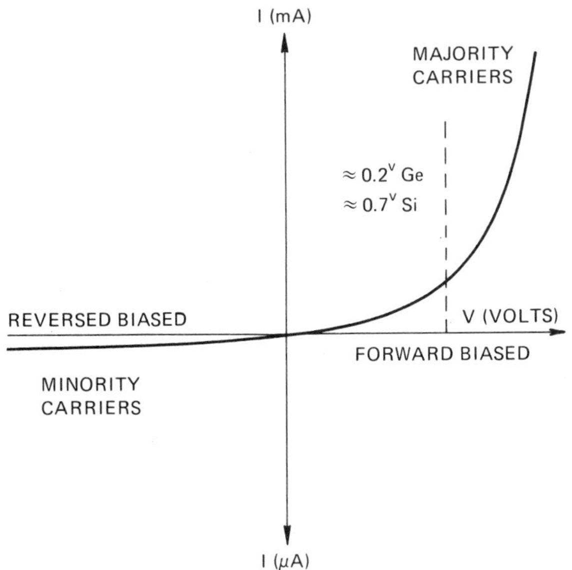

FIG. 1-20 CHARACTERISTIC OF A SEMICONDUCTOR DIODE

In this manner, the barrier around the junction is reduced so that some electrons and holes are able to cross the junction and then recombine. For each electron that recombines, a new electron enters the *n* material from the battery; for each hole that recombines, a new hole forms when an electron leaves the *p* material and flows toward the positive terminal of the battery, figure 1-22.

Since the barrier region is reduced when a *p-n* junction is forward biased, considerable current can flow. A resistor is usually placed in a forward-biased circuit to limit the flow of current so that the junction is not destroyed.

List five techniques for joining p-n junctions. (R1-11)

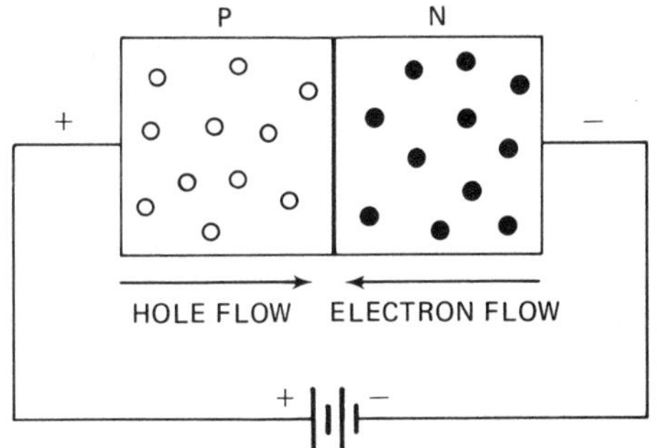

FIG. 1-21 FORWARD-BIASED SEMICONDUCTOR DIODE

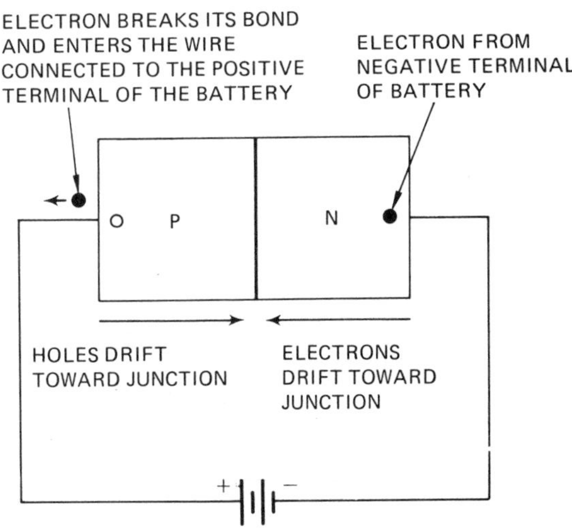

FIG. 1-22 MOVEMENT OF ELECTRONS FOR A FORWARD-BIASED SEMICONDUCTOR DIODE

Explain the action which occurs when a p-n junction is formed. (R1-12)

Define the following terms: (a) donor ions; (b) barrier; (c) depletion region. (R1-13)

Explain the action of a forward-biased p-n junction and that of a reverse-biased p-n junction. (R1-14)

What is the leakage current? (R1-15)

SEMICONDUCTOR DIODE

The electrical symbol for a semiconductor or solid-state diode is shown in figure 1-23. Note that the arrow in the symbol points in the direction of conventional (positive) current flow. Conventional current is the motion of positive charges (holes) "falling downhill" from a positive to a negative potential. Conventional current will be shown as a solid arrow in the diagrams of this text.

Electron flow is the motion of negative charges (electrons) from a negative to a positive potential. Hence, electrons and conventional current flow in opposite directions in the same circuit. Electron flow will be shown as a dotted arrow in the diagrams of this text.

Conventional current is generally used in electronic circuits to conform with the symbolism of diodes and transistors. In other words, the diode symbol in figure 1-23 (a) shows an arrow pointing in the direction of positive charges, or conventional current. Conventional current will be used primarily in this text.

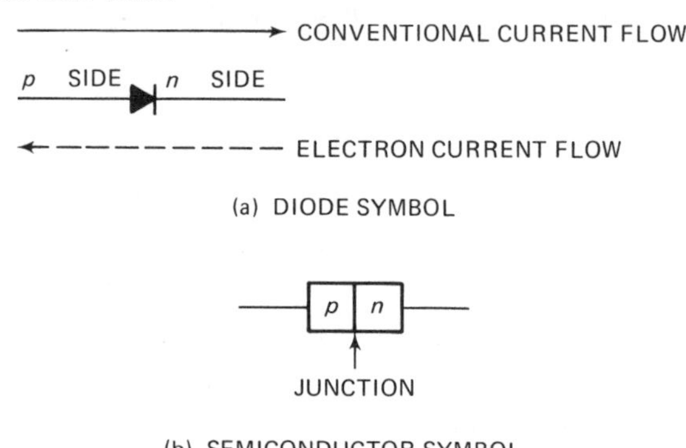

FIG. 1-23 SEMICONDUCTOR DIODE SYMBOL

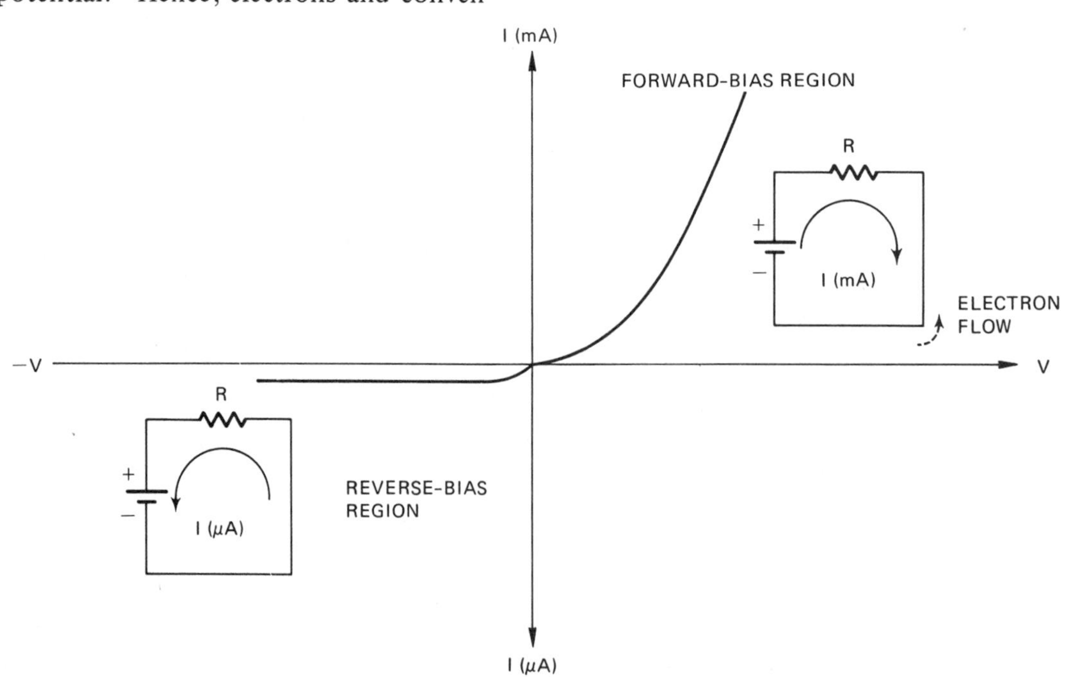

FIG. 1-24 FORWARD-AND REVERSE-BIAS CIRCUITS AND REGIONS OF THE SEMICONDUCTOR DIODE

A representative characteristic curve is shown in figure 1-24, page 16, for semiconductor diodes. These diodes conduct current easily when they are forward biased and have a small amount of leakage current when they are reverse biased.

Diode characteristics depend upon the material used, the doping concentration, and the type of construction used to join the *p-n* junction. When germanium is used as the semiconductor material, a measurable current is detected when the voltage applied in the forward direction is equal to 0.2 to 0.3 volts. When silicon is used, the voltage in the forward direction must be 0.6 to 0.7 volts before a measurable current can be detected. The leakage current is much less for silicon than for germanium.

Why does silicon require a higher forward voltage than germanium to cause a current flow? (R1-16)

LABORATORY EXERCISE 1-1: ANALYSIS OF SEMICONDUCTOR DIODES

PURPOSE

- To determine the *p* and *n* side of a semiconductor using an ohmmeter.
- To observe and graph the effects of diode current for a forward- and reverse-biased semiconductor diode.
- To determine the dc forward and reverse resistance of a semiconductor diode.

MATERIALS

1 Variable dc power supply
1 Vacuum-tube voltmeter (VTVM)
1 Volt-ohm-milliammeter (VOM)
1 Ammeter, 0-100 μA, dc
1 Diode, 1N69
1 Diode, 1N645
1 Diode, 1N2615
1 Diode, HEP170
1 Resistor, 220 Ω, 3W

PROCEDURE

A. Examine the diodes obtained from the instructor and look for the circular band(s) near one end of each of the diodes. The circular band marks the *n* (-) side of the diode (the other side of the diode is the *p* (+) side), figure 1-25.

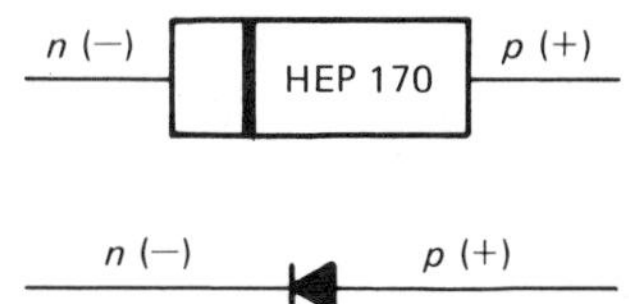

FIG. 1-25 DESIGNATING THE DIODE POLARITIES

B. 1. Use the VOM to check the dc forward and reverse resistances of each of the diodes. To observe the forward dc resistance of the diode, connect an ohmmeter with the positive lead on the *p* side and the negative lead on the *n* side. The ohmmeter will measure a low forward dc resistance. Record this resistance for each diode in Table 1-2.

TABLE 1-2

DIODE	FORWARD RESISTANCE IN Ω	REVERSE RESISTANCE IN MΩ
IN69		
IN645		
IN2615		
HEP170		

2. To observe the reverse dc resistance of a diode, connect an ohmmeter with the negative lead on the *p* side and the positive lead on the *n* side. The ohmmeter will measure a high reverse dc resistance. Record this resistance for each diode in Table 1-2.

C. Connect the circuit shown in figure 1-26 using the IN69 diode.

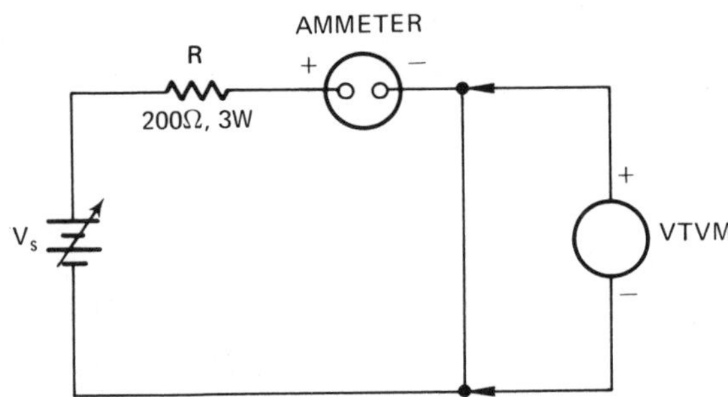

FIG. 1-26 CIRCUIT USED TO OBSERVE THE EFFECT OF FORWARD BIAS ON CURRENT THROUGH A SEMICONDUCTOR DIODE

1. For the forward bias situation, measure and record in Table 1-3 the ammeter readings for the given VTVM voltages.
2. Reverse the diode in the circuit of figure 1-26. Insert the 0-100 μA ammeter. Obtain each of the voltages given in Table 1-4 and record the corresponding current.

   CAUTION: Before recording the currents, remove one lead of the VTVM from the circuit so that the VTVM does not draw current from the power supply.
3. If time permits, other diodes may be inserted in the circuit and steps C.1. and C.2. repeated.
4. From the data obtained, draw a graph (s) of current versus voltage in the forward and reverse directions.

EXTENDED STUDY TOPICS

1. An electron absorbs energy from an external source E_2. As a result, the radius of the electron changes from r_1 to r_2. Derive an equation which shows the increase in total energy ΔE_T in terms of the initial radius r_1 and the final radius r_2.

Table 1-3

VTVM, Volts	I_F, mA
0.00	
0.10	
0.15	
0.20	
0.25	
0.30	
0.35	
0.40	
0.45	
0.50	
0.55	
0.60	
0.65	
0.70	
0.75	

Table 1-4

VTVM, Volts	I_R, μA
0	
1	
2	
3	
4	
5	
8	
10	
15	
20	
25	
30	
40	
45	
50	

2. Initially, the electron in a hydrogen atom is in an orbit with a radius of 5.3×10^{-11} meters. The atom absorbs 2.04×10^{-18} joules of energy from an external source. Find the new radius of the electron.
3. Derive an equation which shows the decrease in total energy ΔE_T in terms of initial radius r_1 and the final radius r_2.
4. Define the following terms:
   a. forbidden gap
   b. hole carriers
   c. intrinsic semiconductor
   d. extrinsic semiconductor
5. What is an ionized atom?
6. Explain the concept of hole flow.
7. May the atom be considered as being made of mostly empty space? Explain.
8. Find the constant A of Eq. 1.11 for (a) germanium and (b) silicon.
9. Find the constant B of Eq. 1.10 for (a) germanium holes and (b) silicon electrons. Find the resistivity for intrinsic silicon at 200° K.

10. Define the following terms:
    a. valence band
    b. energy gap
    c. conduction band
    d. eV
11. How wide is the energy gap for germanium and silicon at (a) room temperature and (b) absolute zero?
12. Define the following terms:
    a. intrinsic germanium
    b. covalent bonding
    c. hole
    d. electron-hole pair

Unit 2

Rectification

OBJECTIVES

After studying this unit, the student will be able to discuss and demonstrate an understanding of the basic principles of:

- The use of a diode as a half-wave rectifier and as a full-wave rectifier.
- Ideal transformer theory
- The advantages offered by a bridge rectifier circuit over the two-diode, full-wave rectifier circuit.
- The calculation of the dc voltage values for half-wave and full-wave rectifiers.

DIODE USED AS A RECTIFIER

It was shown in Unit 1 that the forward-biased semiconductor diode has *low* resistance, and the reverse-biased diode has *large* resistance. In other words, a semiconductor diode conducts current easily in the forward direction but conducts only a small leakage current in the reverse direction.

A diode is used frequently to *rectify* (or change) an ac voltage as shown in figure 2-1, page 22. A sine-wave voltage is applied to a series circuit containing a semiconductor diode and a resistor, figure 2-1a. The diode will conduct only when the *p* region has a positive voltage applied to it relative to the *n* region. A positive voltage is applied during the time interval between 0 and π, figure 2-1b. The diode will conduct during this interval and a resulting voltage drop is developed across the resistor R. During the time interval π to 2π, the applied voltage causes the *p* region to be negative relative to the *n* region with the result that the diode is reverse biased. There is very little current flow in this case and no voltage will be developed across the resistor R. An output voltage will appear across R as frequently as the input voltage v_{in} repeats, figure 2-1c. In other words, the output voltage has the same frequency as the input voltage. At this point, it is assumed that the forward voltage drops of the diode are negligible. The output voltage across R actually

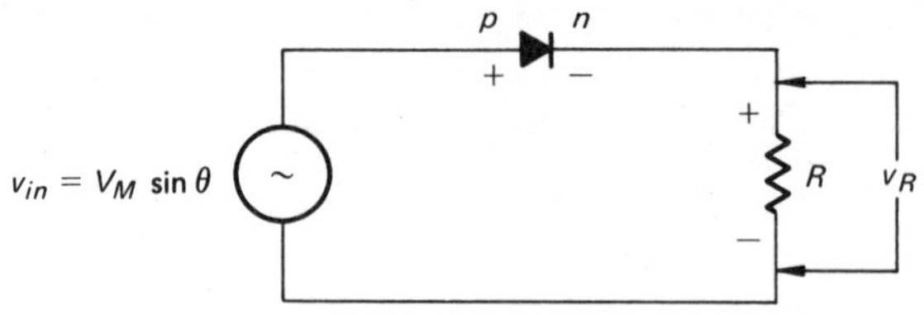

(a) HALF-WAVE RECTIFIER CIRCUIT

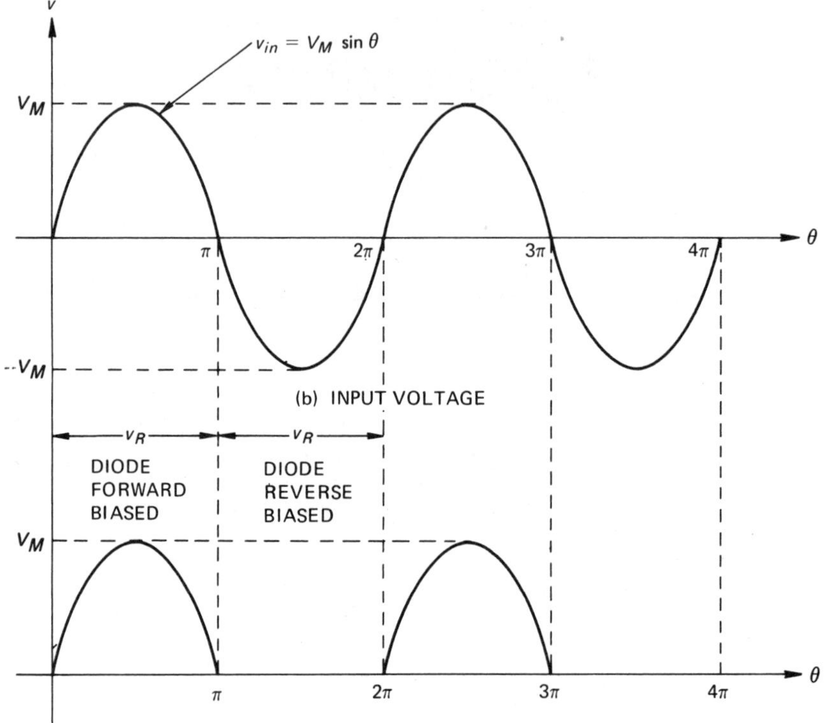

(c) VOLTAGE ACROSS RESISTOR R (v_R)

FIG. 2-1 HALF-WAVE RECTIFIER CIRCUIT AND VOLTAGE WAVEFORMS

is the difference between the input voltage and the diode forward voltage drop when the diode conducts.

If the diode in figure 2-1a is turned around, it will conduct during the negative portion of the input signal. The diode will be reverse biased during the positive portion of the input signal, figure 2-2.

The *average* or *dc value* of the half-wave rectified sine-wave voltage across the resistor is shown in figure 2-3. The dc value is given by the equation:

$$V_{dc\,(hw)} = 0.318\ V_M \qquad \text{Eq. 2.1}$$

where V_M = the maximum or peak value of the sine wave

$V_{dc\,(hw)}$ = the average or dc value of the half-wave rectified sine wave

There are two important diode parameters that must not be exceeded if the diode is to operate properly. These parameters are the *repetitive peak reverse voltage (PRV)* and the *maximum allowable dc output current.*

As shown in figure 2-4, there will be a reverse voltage across the diode when it is not conducting. The repetitive peak reverse volt-

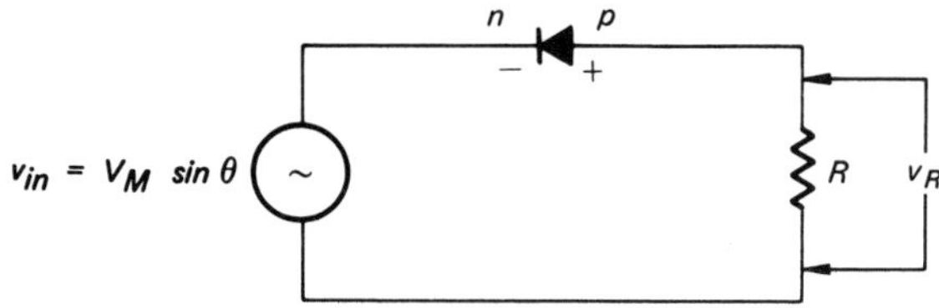

(a) HALF-WAVE RECTIFIER CIRCUIT

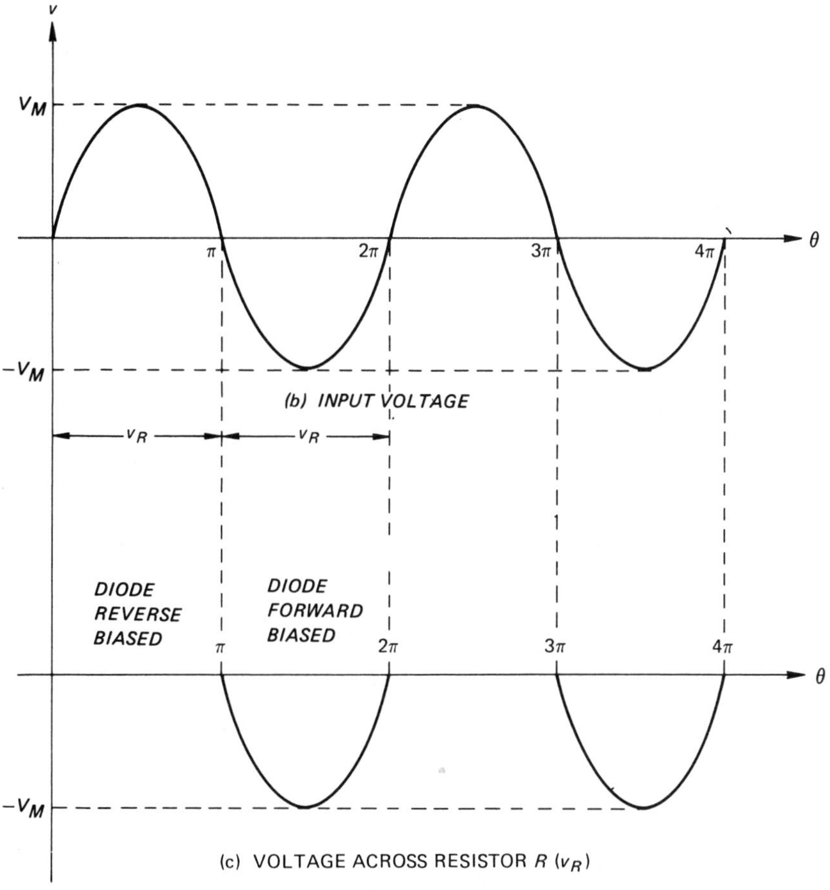

FIG. 2-2 HALF-WAVE RECTIFIER CIRCUIT AND VOLTAGE WAVEFORMS

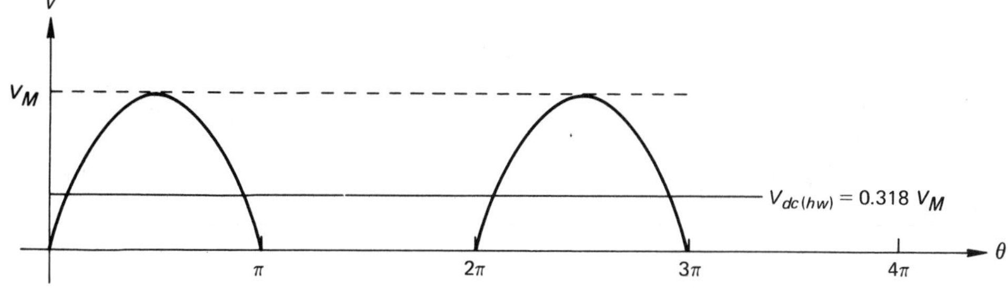

FIG. 2-3 DC VALUE OF A HALF-WAVE RECTIFIED SINE WAVE

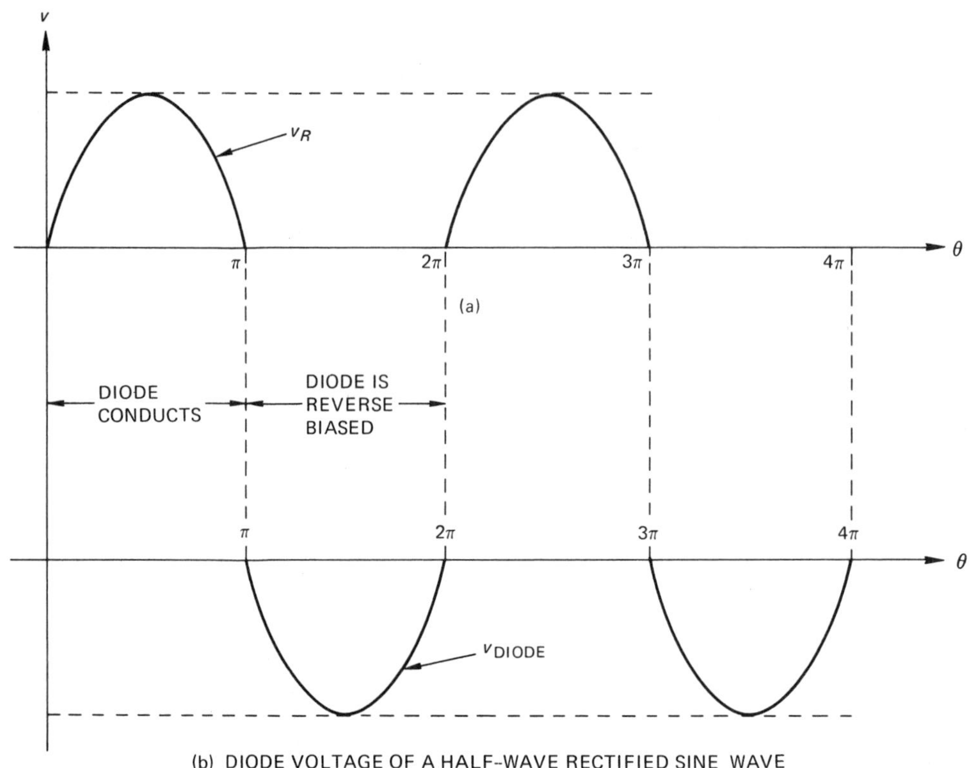

(b) DIODE VOLTAGE OF A HALF-WAVE RECTIFIED SINE WAVE

FIG. 2-4 OUTPUT AND DIODE VOLTAGE OF A HALF-WAVE RECTIFIER

age (PRV) rating for a diode indicates the highest value of reverse voltage that can be applied repeatedly to the diode without causing harmful effects due to a condition called avalanche (this condition is covered in Unit 4). The PRV rating must be equal to or greater than the peak value of the sine wave, V_M.

The maximum allowable dc output current is always specified for a conducting diode to indicate the power capabilities of the diode. This current is the maximum operating current that will keep the *p-n* junction within its allowed temperature at a specific ambient reference temperature. Ambient temperature is the temperature surrounding the outer case of a diode; the ambient temperature may be the room temperature.

PROBLEM 1.

Using the half-wave rectifier circuit in figure 2-1a, find

a. the voltage across R as measured with a dc voltmeter,

b. the current in series with R (using a dc ammeter), and

c. the minimum PRV rating of the diode.

For the circuit, V_M = 314 V and R = 1 kΩ.

a. $V_{dc\,(hw)} = 0.318\,V_M = 0.318 \times 314\ V = 100\ V$

b. $I_{dc} = \dfrac{V_{dc}}{R} = \dfrac{100\ V}{1\ k\Omega} = 100\ mA$

c. $PRV = V_M = 314\ V$

For the half-wave rectifier circuit of figure 2-1, V_M = 400 V and R = 2.2 kΩ. Find (a) the dc voltage across R, (b) the minimum PRV rating of the diode, and (c) the maximum allowable dc output current of the diode. (R2-1)

A half-wave rectifier circuit develops a dc voltage of 100 V across a resistor of 200 Ω. Find the minimum PRV rating of the diode. (R2-2)

A half-wave rectifier circuit has a dc current of 100 mA through a 560 Ω load resistance. Find (a) the dc output voltage and (b) the minimum PRV rating of the diode. (R2-3)

IDEAL TRANSFORMER THEORY

Figure 2-5 illustrates an iron core transformer. The transformer consists of two windings of wire wrapped about the iron core. The two windings are electrically isolated from each other. The winding associated with the source or input is called the *primary turns* winding and the number of turns is designated by N_P; the second (output) winding is called the *secondary turns* winding and the number of turns in this winding is designated by N_S. The secondary winding is usually connected to a load.

When the primary winding is connected to an ac voltage source, an ac current will flow through the primary. The ac current causes an ac magnetic field (flux) that flows through the core linking the secondary winding to the primary winding. The magnetic flux passing through the secondary winding causes an in-

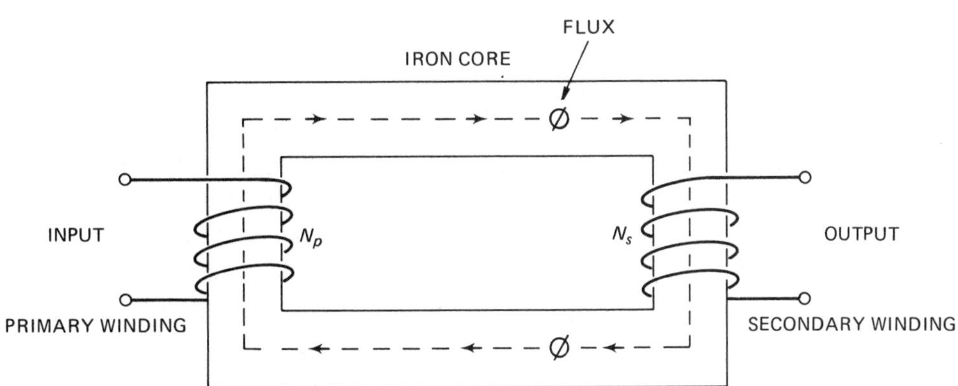

(a) BASIC IDEAL TRANSFORMER

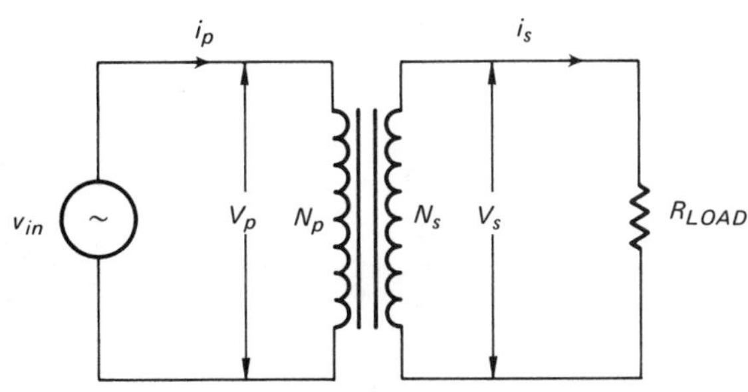

(b) SCHEMATIC REPRESENTATION OF THE TRANSFORMER

FIG. 2-5 IDEAL TRANSFORMER REPRESENTATION

duced emf according to Faraday's law of mutual induction.

In the case of an *ideal* transformer, all of the flux produced in the primary winding will link the secondary winding to the primary. Because the same flux cuts each conductor wound on the core, the induced emf will be the same for each turn; therefore, the voltage across each winding will be proportional to the number of turns. The following equation expresses this relationship.

$$\frac{V_p}{V_s} = \frac{N_p}{N_s} \qquad \text{Eq. 2.2}$$

where V_p = voltage across primary winding

V_s = voltage across secondary winding

N_p = primary turns

N_s = secondary turns

A transformation ratio (turns ratio) is defined as:

$$a = \frac{N_p}{N_s} \qquad \text{Eq. 2.3}$$

From Eq. 2.2, it can be seen that the secondary voltage can be raised or lowered by changing the ratio of the turns. That is, if the number of turns on the secondary winding is increased, then the secondary voltage increases.

If a load is added to the secondary winding of the transformer, the emf induced by the primary flux will cause current to flow in the secondary circuit. The current will cause an mmf of $N_S I_S$ ampere-turns to appear across the secondary. Since the secondary ampere-turns are due to the primary, then the secondary and the primary ampere-turns ($N_P I_P$) are the same (assuming ideal conditions). In other words,

$$N_p I_p = N_s I_s$$

This equation also can be written as:

$$\frac{I_P}{I_S} = \frac{N_S}{N_P} = \frac{1}{a} \qquad \text{Eq. 2.4}$$

where I_p = primary current

I_s = secondary current

The power delivered to the primary must equal the power of the secondary for an ideal transformer.

$$P_p = P_s$$

$$I^2{}_p R_p = I^2{}_s R_s$$

$$\frac{R_p}{R_s} = \frac{I^2{}_s}{I^2{}_p} = a^2$$

Therefore,

$$R_p = a^2 R_s \qquad \text{Eq. 2.5}$$

where R_p = reflected resistance into primary

R_s = resistance across secondary

Some transformers are provided with a center tap on the secondary winding, as shown in figure 2-6, to increase the number of available output voltages. From the transformer voltage ratings shown in the figure, it can be seen that the primary voltage is 120 V, the secondary voltage from the top terminal to the center tap is 250 V, and the voltage from the top terminal to the bottom terminal (the voltage across the entire secondary) is 500 V.

Transformers are used in rectifier circuits to step up or step down the primary voltage so that the secondary voltage(s) will satisfy the voltage requirements for particular applications. Most transformers are provided with a

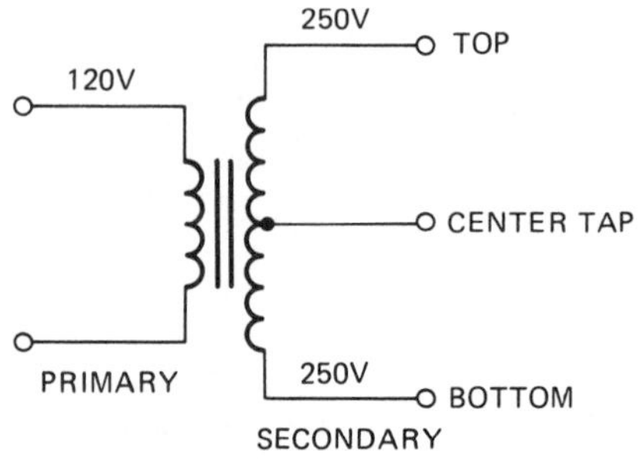

FIG. 2-6 TRANSFORMER WITH A CENTER TAP

120-V primary to permit the transformer to be connected to a standard wall outlet.

Another reason for the use of transformers in rectifier circuits is to electrically isolate the secondary from the primary. In other words, the electrical circuits connected to the secondary will not affect the primary, and the circuits connected to the primary will not affect the secondary.

PROBLEM 2

A transformer has a turns ratio of 20 and a primary voltage of 120 V. If a 2-kΩ resistor is connected to the secondary, find:

a. V_s,
b. I_s,
c. I_p, and
d. R_p.

a. $\frac{V_p}{V_s} = a$

$$V_s = \frac{V_p}{a} = \frac{120\text{ V}}{20}$$

$= \mathbf{6\ V}$

b. $I_s = \frac{V_s}{R_s} = \frac{6\text{ V}}{2\text{ k}\Omega}$

$= \mathbf{3\ mA}$

c. $I_p = \frac{I_s}{a} = \frac{3\text{ mA}}{20}$

$= \mathbf{0.15\ mA}$

d. $R_p = a^2\ R_s$
$= (20)^2 \times 2\text{ k}\Omega$
$= \mathbf{800\ k\Omega}$

A transformer has a turns ratio of 1/20, a primary voltage of 120 V, and a 2 kΩ load connected to the secondary. Find V_s, I_s, I_p, and R_p. (R2-4)

A transformer has a turns ratio of 12, V_p = 40 V, and R_s = 500 Ω. Find V_s, I_s, I_p, and R_p. (R2-5)

Repeat review question (R2-5) for a turns ratio of 1/30. (R2-6)

TWO-DIODE FULL-WAVE RECTIFIER

The two-diode full-wave rectifier circuit shown in figure 2-7a requires two ac voltage inputs 180° out of phase with respect to the common point (0). The voltage polarities of the ac sources are shown in figure 2-7a for the interval between 0 and π of the ac sine wave. The polarities of the sources will reverse in the interval between π and 2π. The ac sine wave representations of the two sources v_{01} and v_{02} are shown in figures 2-7b and 2-7c respectively. To analyze the two-diode, full-wave rectifier circuit, it is assumed that the diodes are ideal; that is, there is no voltage drop across the diodes when they conduct and no leakage current when the diodes are reverse biased.

During the positive half-cycle of v_{01} (+A), diode D_1 conducts since its *p* side is positive relative to its *n* side. (D_1 is forward biased). At the same time, the negative half-cycle (-B) of v_{02} is applied to the *p* side of diode D_2. In other words, D_2 is reverse biased and no current flows through this diode. Therefore, for the ac sine-wave interval from 0 to π, the total effect is that of ac current flowing from voltage source v_{01}, through diode D_1, and back through resistor R_L as shown in figure 2-8. Remember that diode D_2 is reverse biased and is not conducting.

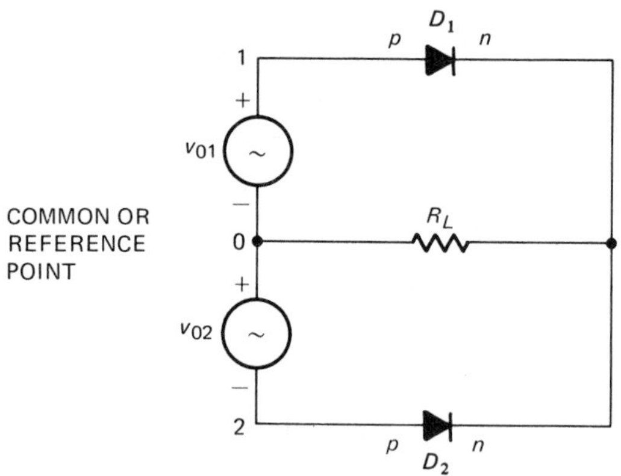

(a) TWO-DIODE FULL-WAVE RECTIFIER CIRCUIT

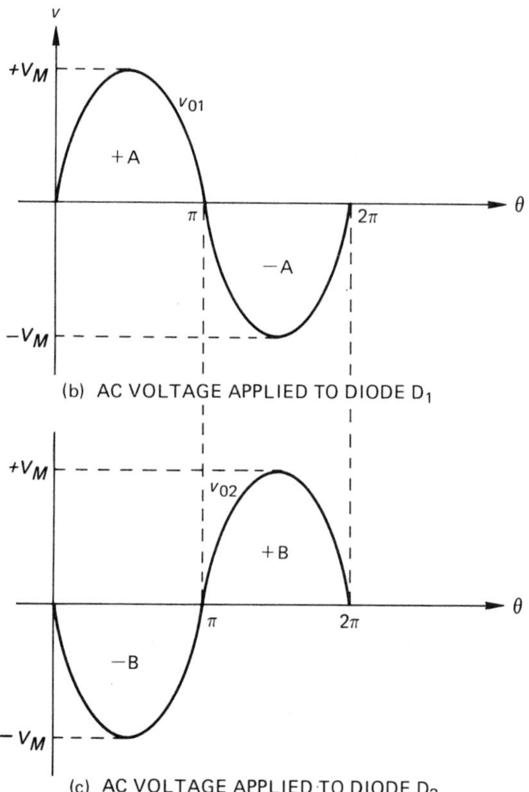

(b) AC VOLTAGE APPLIED TO DIODE D_1

(c) AC VOLTAGE APPLIED TO DIODE D_2

FIG. 2-7 TWO-DIODE FULL-WAVE RECTIFIER CIRCUIT AND THE AC INPUT VOLTAGES

For the interval between π and 2π the polarities of the source voltages reverse as shown in figure 2-9. (Compare the polarities in figures 2-8 and 2-9.) When the negative half-cycle (-A) of v_{01} is applied to diode D_1, the diode will be reverse biased since its *p* side is more negative than its *n* side. At the same time, the positive half-cycle (+B) of v_{02} is applied to the *p* side of diode D_2. As a result, diode D_2 conducts during the interval between π and 2π.

For this interval, therefore, the total effect is that of an ac current flowing from the voltage source v_{02}, through diode D_2, and back through resistor R_L as shown in figure 2-9. Remember that diode D_1 is reverse biased and is not conducting at this point.

The total voltage across R_L is equal to the voltage drop $[i_{(+A)}\ R_L]$ caused by the conduction of diode D_1, plus the voltage drop $[i_{(+B)}\ R_L]$ caused by the conduction of diode D_2. The entire voltage wave of R_L (v_R) is shown in figure 2-10. This waveshape is called a full-wave rectified sine wave. The output voltage across R_L repeats twice as often as the positive input sine wave voltage. In other words, the output voltage of a full-wave rectifier circuit has a frequency equal to twice that of the input frequency of the sine wave.

A diode in the two-diode full-wave rectifier circuit must be able to handle a minimum peak reverse voltage (PRV) of $2V_M$. In figure 2-8, if Kirchhoff's voltage law is applied to the bottom portion of the circuit containing the reverse-biased diode D_2, $v_{D2} = v_{02} + v_R$. When the ac source voltages are at their maximum values, $V_{D2} = V_M + V_M = 2V_M$. In other words, the PRV value of the diode must be at least twice the maximum value of the ac sine wave.

The average or dc value of the voltage for a full-wave rectifier circuit is expressed by the following equation:

$$V_{dc\,(fw)} = 0.636\ V_M \qquad \text{Eq. 2.6}$$

Note that Eq. 2.6 is equal to twice the average or dc voltage for a half-wave rectifier circuit, Eq. 2.1

Practical two-diode full-wave rectifier circuits require the use of a center-tapped transformer to produce the ac voltages

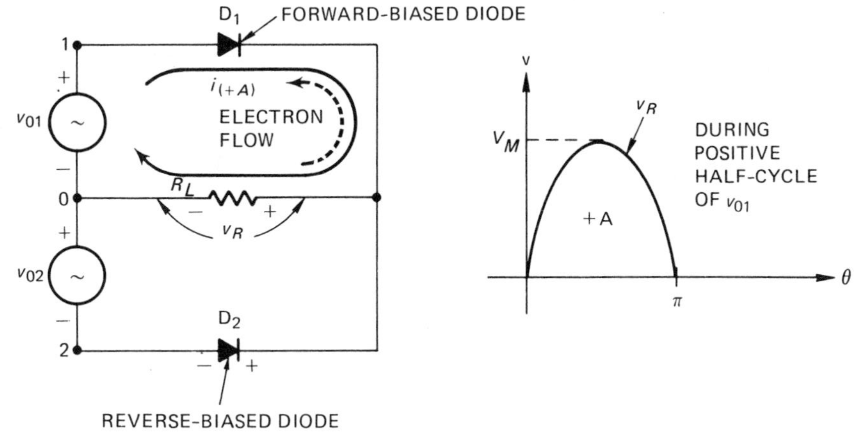

FIG. 2-8 CURRENT FLOW THROUGH R_L WHEN DIODE D_1 CONDUCTS

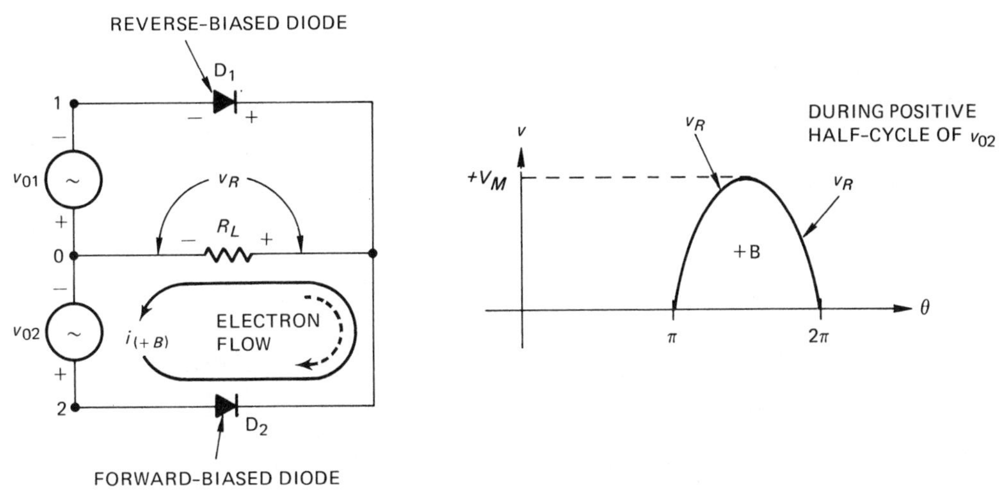

FIG. 2-9 CURRENT FLOW THROUGH R_L WHEN DIODE D_2 CONDUCTS

necessary for the proper operation of the circuit. The center-tapped transformer usually has a 120-V primary winding as shown in figure 2-11, page 30. The secondary voltage must develop a peak voltage of V_M across each half of the transformer.

PROBLEM 3.

For the half-wave rectifier circuit in figure 2-11, V_p = 120 V (rms), V_s = 141.4 V (rms), and R_L = 10 kΩ. Find a. the voltage a dc voltmeter will read across R_L, b. the current a dc ammeter will read in series with R_L, and c. the minimum PRV rating of the diodes.

a. $V_{dc\,(fw)} = 0.636\ V_M$

$V_M = 1.414 \times V_s$

$= 1.414 \times 141.4\ \text{V}$

$= 200\ \text{V}$

$V_{dc\,(fw)} = 0.636 \times 200\ \text{V}$

$= \mathbf{127.2\ V}$

b. $I_{dc} = V_{dc\,(out)}/R_L$

$= \dfrac{127.2\ \text{V}}{10\ \text{k}\Omega}$

$= \mathbf{12.72\ mA}$

c. PRV $= 2\ V_M$

$= 2 \times 200\ \text{V}$

$= \mathbf{400\ V}$

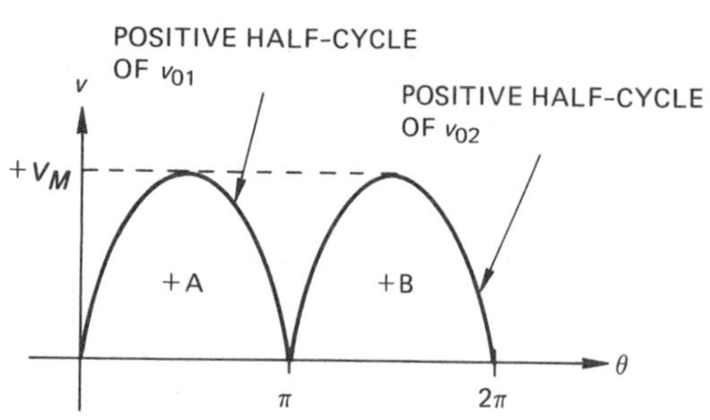

FIG. 2-10 THE VOLTAGE ACROSS R_L (v_R) FOR A FULL-WAVE RECTIFIER

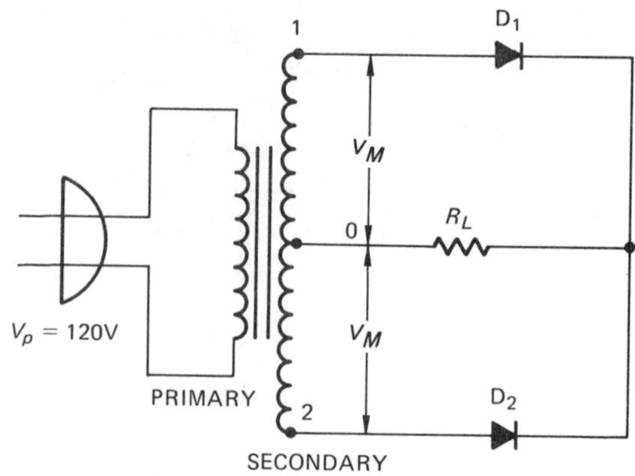

FIG. 2-11 CENTER-TAPPED TRANSFORMER IN A TWO-DIODE FULL-WAVE RECTIFIER CIRCUIT

For the full-wave rectifier circuit of figure 2-11, V_p = 120 V (rms), V_s = 250 V (rms), and R_L = 1 k Ω. Find the dc voltage across R_L, the dc current through R_L, and the minimum PRV rating of the diodes. (R2-7)

For the full-wave rectifier circuit of figure 2-11, the transformer turns ratio is 0.05, V_p = 120 V (rms), and the dc output current is 100 mA. Find the dc output voltage, the primary transformer current, the reflected resistance R_p, and the minimum PRV rating of the diode. (R2-8)

Draw the schematic diagram for a two-diode full-wave rectifier circuit with negative going voltage cycles. (R2-9)

FULL-WAVE BRIDGE RECTIFIER CIRCUIT

The bridge rectifier circuit shown in figure 2-12 gives the same results as the two-diode full-wave rectifier circuit. The bridge circuit requires four diodes instead of two, but the transformer does not require a center tap. The PRV rating of the diodes in the bridge circuit is V_M, rather than the value of the diodes in the two-diode circuit, $2V_M$.

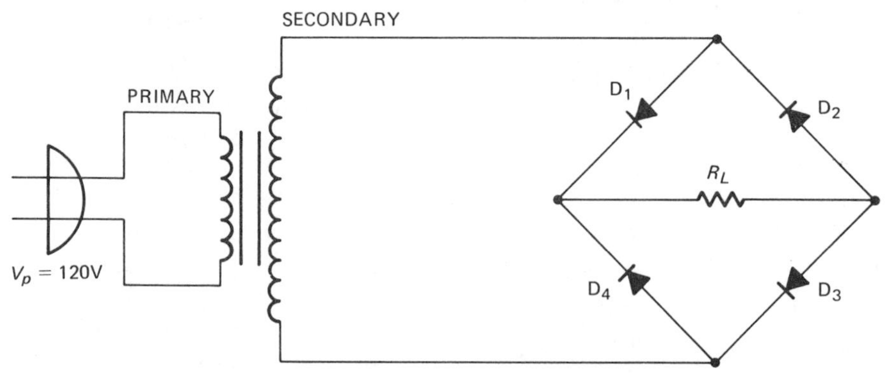

FIG. 2-12 FULL-WAVE BRIDGE RECTIFIER CIRCUIT

The voltage polarities supplied by an ac voltage applied to the secondary of the transformer determine the circuit operation. During the positive half-cycle of the sine wave, the voltage polarity is as shown in figure 2-13. Diodes D_1 and D_3 are forward biased and are conducting since their *p* sides are more positive relative to their *n* sides. At the same time, diodes D_2 and D_4 are reverse biased since their *n* sides experience negative voltage relative to their *p* sides. The current through R_L is shown in figure 2-13a.

During the negative half-cycle of the sine wave, the voltage polarity is as shown in figure 2-14. Diodes D_2 and D_4 are forward biased and are conducting since their *p* sides are more positive relative to their *n* sides. Diodes D_1 and D_3 are reverse biased since their *n* sides experience negative voltage relative to their *p* sides. The current through R_L is shown in figure 2-14a.

The current direction through R_L is the same for both figures 2-13 and 2-14, page 32. In other words, the voltage polarity across R_L does not change for either half-cycle of the sine wave. As a result, the voltage waveshape across R_L is identical to the waveshape obtained with the full-wave rectifier circuit containing two diodes.

The circuit in figure 2-15, page 32, shows that when diodes D_2 and D_4 are reverse biased, they are forced by the conducting diodes D_1 and D_3 to be in parallel with R_L. This situation results because diodes D_1 and D_3 are short

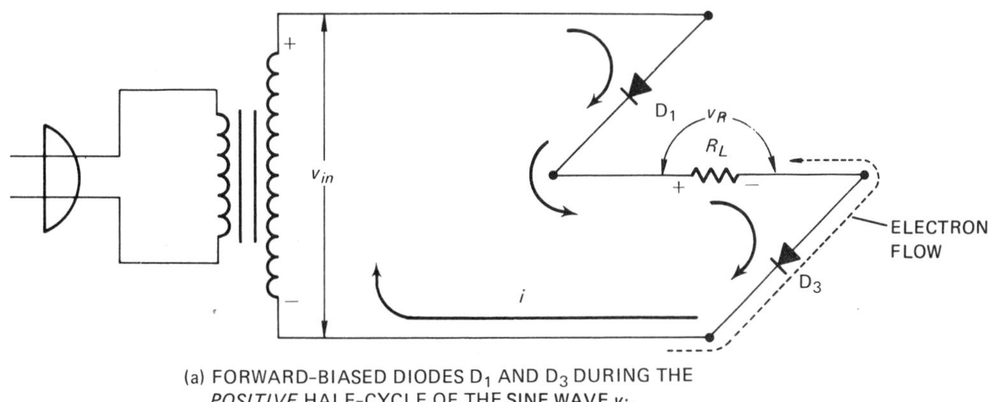

(a) FORWARD-BIASED DIODES D_1 AND D_3 DURING THE *POSITIVE* HALF-CYCLE OF THE SINE WAVE v_{in}

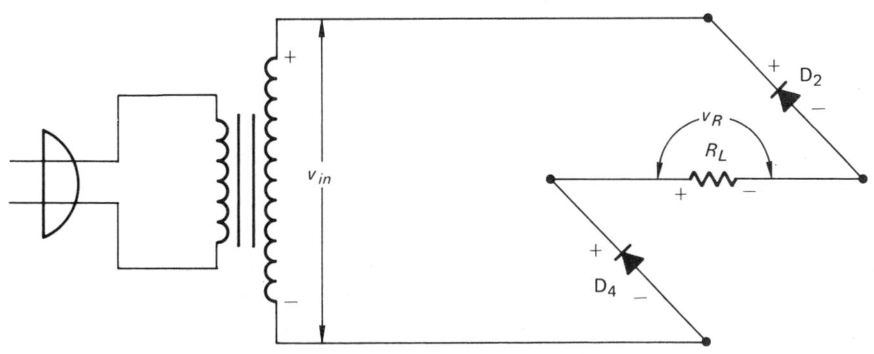

(b) REVERSE-BIASED DIODES D_2 AND D_4 DURING THE *POSITIVE* HALF-CYCLE OF THE SINE WAVE v_{in}

FIG. 2-13 BRIDGE RECTIFIER CIRCUIT DURING THE POSITIVE HALF-CYCLE OF THE SINE WAVE

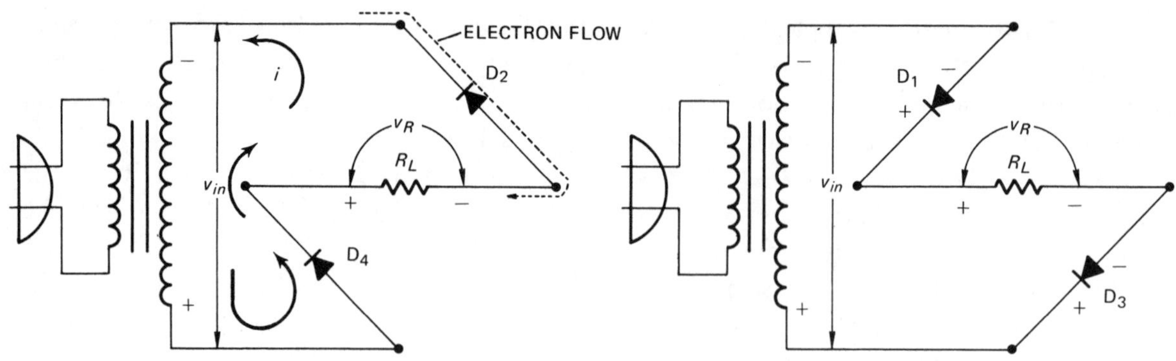

(a) FORWARD-BIASED DIODES D_2 AND D_4 DURING THE *NEGATIVE* HALF-CYCLE OF THE SINE WAVE v_{in}

(b) REVERSE-BIASED DIODES D_1 AND D_3 DURING THE *NEGATIVE* HALF-CYCLE OF THE SINE WAVE v_{in}

FIG. 2-14 BRIDGE RECTIFIER CIRCUIT DURING THE NEGATIVE HALF-CYCLE OF THE SINE WAVE

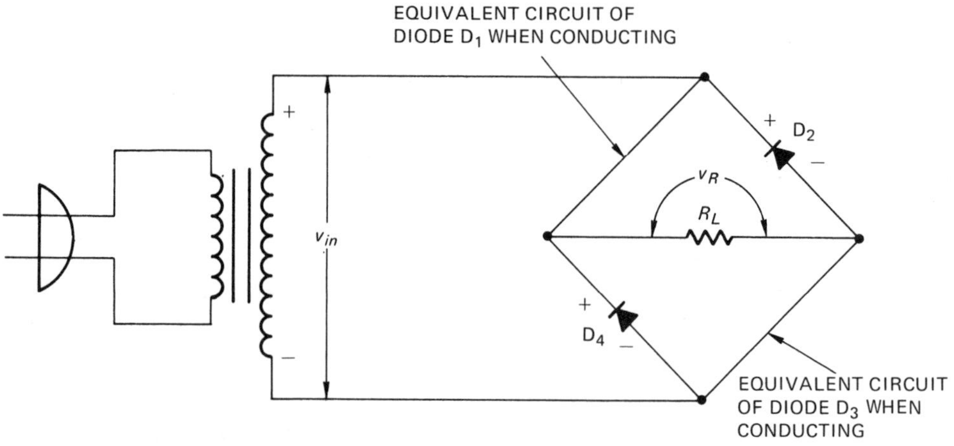

FIG. 2-15 REVERSE-BIASED DIODES D_2 AND D_4 SHOWN IN PARALLEL WHEN DIODES D_1 AND D_3 CONDUCT

circuit paths when they conduct. As a result, the maximum reverse voltage that can exist across diodes D_2 and D_4 is V_M. The full-wave bridge rectifier circuit is more popular than the two-diode full-wave rectifier circuit because of the lower PRV rating required for the diodes in the circuit. The dc voltage across R_L in the bridge circuit is identical to the dc voltage across R_L in the two-diode full-wave rectifier circuit.

Calculate the PRV rating for a bridge rectifier circuit whose dc voltage is 75 V. (R2-10)

A bridge rectifier circuit contains a transformer with a turns ratio of 1/10 and V_p = 120 V. If the load resistance is 10 kΩ, find V_s, $V_{dc(fw)}$, the minimum PRV rating of the diodes, and I_{dc}. (R2-11)

Draw the schematic diagram for a bridge rectifier circuit with negative going voltage cycles. (R2-12)

LABORATORY EXERCISE 2-1:
HALF-WAVE RECTIFIER CIRCUIT

PURPOSE

- To observe the voltage waveshapes associated with a simple half-wave rectifier.
- To measure the dc and ac voltages for a simple half-wave rectifier.
- To calculate the dc current through the resistive load of a simple half-wave rectifier.

MATERIALS

1 Oscilloscope
1 Voltage transformer, stepdown, Triad No. F-54X, primary 117 V, secondary 35 V with center tap (CT) (or equivalent)
1 Vacuum-tube voltmeter (VTVM)
1 Power source, ac, 117 V, 60 Hz.
1 Diode, HEP170 recommended
1 Resistor, 10 kΩ, 1 W

PROCEDURE

A. Connect the circuit as shown in figure 2-16.

B. 1. Using the oscilloscope, measure the voltages V_p, V_s, V_D, and V_L.

CAUTION: Use a two-prong plug or isolated supply for the oscilloscope.

2. Draw the voltage waveforms on graph paper (use paper with a grid of four squares per inch).

3. Record the peak value of all of the voltages.

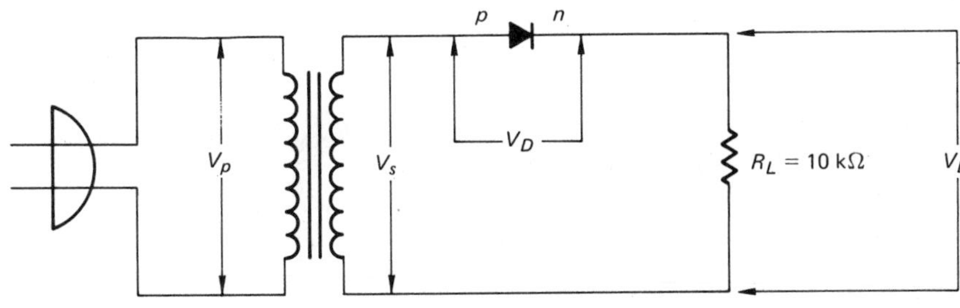

FIG. 2-16 HALF-WAVE RECTIFIER CIRCUIT

C. Using a dc voltmeter, measure the voltages V_p, V_s, V_D, and V_L and record these values in Table 2-1.

D. Using an ac voltmeter, measure the voltages V_p, V_s, V_D, and V_L and record these values in Table 2-1.

Table 2-1

Voltages	Dc reading in volts	Ac reading in volts
V_p		
V_s		
V_D		
V_L		

E. 1. Using the data from step B.1., compute the value of the dc voltage which should appear across R_L.

2. Compute the percent difference between the calculated and measured values of the dc voltage across R_L.

F. Why is there a difference between the ac voltages V_s and V_L?

G. Calculate the dc value of the current (I_L) through R_L.

LABORATORY EXERCISE 2-2: TRANSFORMER

PURPOSE

- To determine the turns ratio of a transformer
- To observe the effect of varying secondary current on primary current
- To measure the voltage from the center tap of the transformer to both sides of the secondary windings
- To learn the procedure for resistance testing a transformer

MATERIALS

1 Voltage transformer, stepdown, Triad No. F-54X, primary 117 V, secondary 35 V CT
1 Vacuum-tube voltmeter (VTVM)
1 Milliammeter, ac, ranges: 0-500 mA and 0-100 mA
1 Power source, variable ac, 117 V, 60 Hz
1 Resistor, 5 Ω, 10 W
1 Resistor, 10 Ω, 10 W
2 Resistors, 33 Ω, 5 W
1 Resistor, 47 Ω, 5 W

PROCEDURE

A. Connect the circuit shown in figure 2-17.

CAUTION: Be sure that the variable ac voltage source is set at 0 V and is connected to the *primary* of the stepdown voltage transformer. The secondary of the transformer must be *open*.

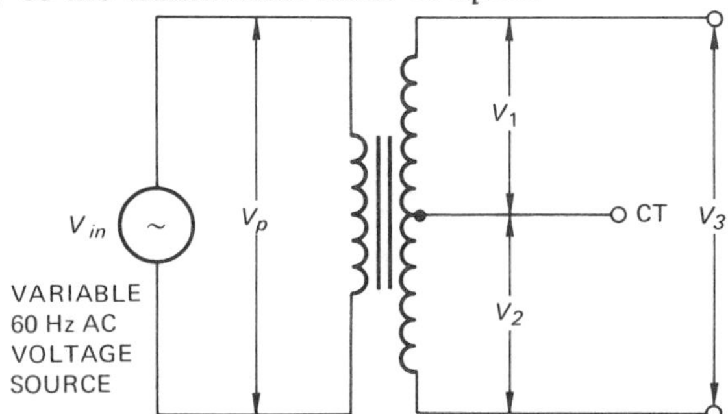

FIG. 2-17 VOLTAGE TRANSFORMER WITH CENTER TAP (CT)

B. 1. Increase the primary voltage (V_p) to 110 V.

2. Measure the ac voltages V_p, V_1, V_2, and V_3. Record these voltage values in Table 2-2.

3. Does the sum of V_1 and V_2 equal V_3? Why?

4. Why is the transformer called a stepdown transformer?

C. Calculate the turns ratio a and record this value in Table 2-2.

Table 2-2

V_p	V_1	V_2	V_3	a
110 V				
50 V				

D. 1. Reduce the primary voltage to 50 V, and repeat steps B.1 and B.2.

2. Calculate the turns ratio from the data in step D.1.

3. Is there any difference between the turns ratios calculated in steps C and D.2.? Why?

E. Reduce the primary voltage of the transformer to 0 V.

F. Connect the circuit as shown in figure 2-18. The ammeter should be set on the 0-500 mA range.

G. 1. Increase the source voltage V_s until the *secondary* voltage is 6 V.

2. Measure the primary current and record this value in Table 2-3, page 36.

H. 1. Repeat steps G.1 and G.2 for the other load resistors shown in Table 2-3.

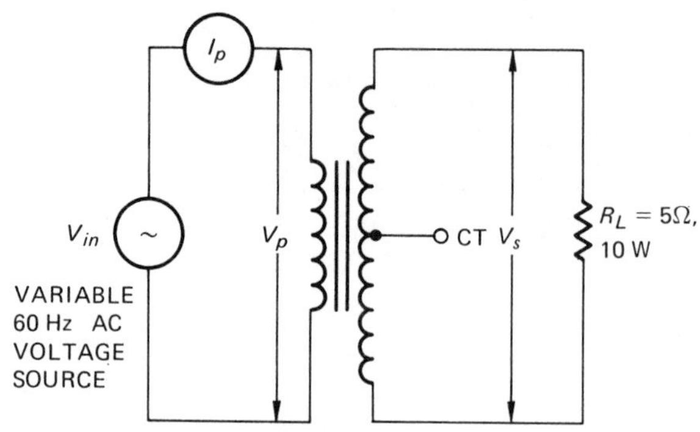

FIG. 2-18 TRANSFORMER WITH A LOAD ON THE SECONDARY

Table 2-3

R_L in ohms	I_p in mA	V_p	V_s in V	$I_s = \frac{V_s}{R_L}$ in mA	$I_p = \frac{I_s}{a}$ in mA	P_s in W
5			6			
10			6			
16.5 Ω by paralleling two 33-Ω resistors			6			
33			6			
47			6			

CAUTION: Before removing each load resistor from the circuit, reduce the ac power source to 0 V.

2. Compute the secondary current I_s for each load and record in Table 2-3.
3. Compute the primary current $I_p = I_s / a$ for each load and record in Table 2-3.
4. Compute the power dissipated (P_s) for each load and record in Table 2-3.
5. Compare the results of steps H.1 and H.3.
6. What is the primary power corresponding to the powers calculated in step H.4.? Why?

I. Reduce the primary voltage to 0 V and disconnect the transformer.

J. Using an ohmmeter, measure the resistances indicated in Table 2-4. Record these resistance values in Table 2-4.

K. Is there any electrical connection between the primary and secondary of the transformer? Why?

Table 2-4

All resistance values in ohms				
Primary	Secondary	Primary to Secondary	Primary to Frame	Secondary to Frame

LABORATORY EXERCISE 2-3:
FULL-WAVE TWO-DIODE RECTIFIER CIRCUIT

PURPOSE

- To observe the voltage waveshapes associated with a full-wave rectifier circuit using two diodes.
- To measure the dc and ac voltages for a two-diode full-wave rectifier circuit.
- To calculate the dc current through the resistive load of a full-wave rectifier circuit.

MATERIALS

1 Oscilloscope
1 Voltage transformer, stepdown, Triad No. F-54X, primary 117 V, secondary 35 V CT
1 Vacuum-tube voltmeter (VTVM)
1 Power source, ac, 117 V, 60 Hz
2 Diodes, HEP170 suggested
1 Resistor, 10 kΩ, 1 W

PROCEDURE

A. Connect the circuit as shown in figure 2-19.

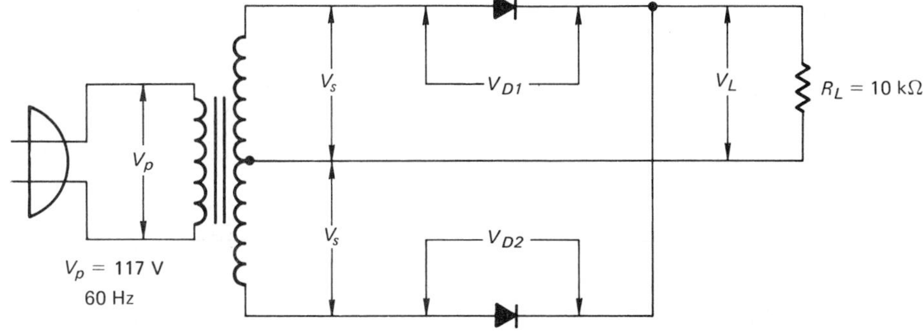

FIG. 2-19 FULL-WAVE RECTIFIER CIRCUIT

B. 1. Using the oscilloscope, measure the voltages V_p, V_s, V_{D1}, V_{D2}, and V_L.

CAUTION: Use a two-prong plug or isolated ac supply for the oscilloscope.

2. Draw the voltage waveforms on graph paper (use paper with a grid of four squares per inch).

3. Record the peak value of each of the voltages.

C. 1. Using a dc voltmeter, measure the voltages V_p, V_s, V_{D1}, V_{D2}, and V_L.

2. Record these values in Table 2-5.

D. 1. Using an ac voltmeter, measure the voltages V_p, V_s, V_{D_1}, V_{D_2}, and V_L.

2. Record these values in Table 2-5.

TABLE 2-5

Voltages	dc reading in volts	ac reading in volts
V_p		
V_s		
V_{D_1}		
V_{D_2}		
V_L		

E. 1. Using the data from step B.1., compute the value of the dc voltage which appears across R_L.

2. Compute the percent difference between the calculated and measured values of the dc voltage across R_L.

F. Why is there a difference between the ac voltages V_s and V_L?

G. Calculate the dc value of current (I_L) through R_L.

LABORATORY EXERCISE 2-4: FULL-WAVE BRIDGE RECTIFIER CIRCUIT

PURPOSE

- To observe the voltage waveshapes associated with a full-wave bridge rectifier circuit.
- To measure the dc and ac voltages for a full-wave bridge rectifier circuit.
- To calculate the dc current through the resistive load of a full-wave rectifier circuit.

MATERIALS

1 Oscilloscope
1 Voltage transformer, stepdown, Triad No. F-54X, primary 117 V, secondary 35 V CT
1 Vacuum-tube voltmeter (VTVM)
1 Power source, ac, 117 V, 60 Hz
4 Diodes, HEP170 suggested
1 Resistor, 10 kΩ, 1 W

PROCEDURE

A. Connect the circuit as shown in figure 2-20.

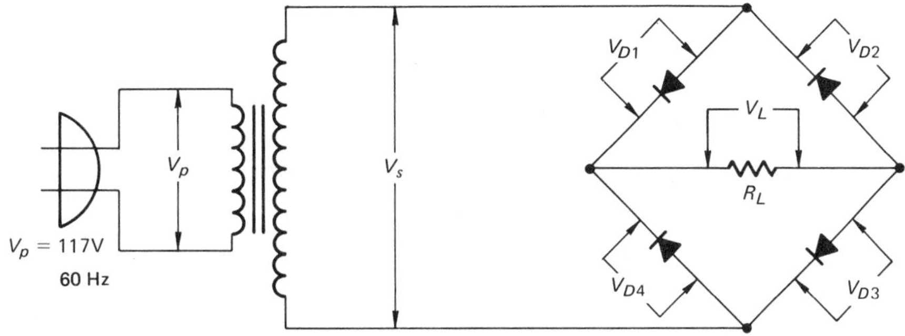

FIG. 2-20 FULL-WAVE RECTIFIER BRIDGE CIRCUIT

B. 1. Using the oscilloscope, measure the voltages V_p, V_s, V_{D1}, V_{D2}, V_{D3}, V_{D4}, and V_L.

CAUTION: Use a two-prong plug or isolated ac supply for the oscilloscope.

2. Draw the voltages on graph paper (use paper with a grid of four squares per inch).

3. Record the peak value of each of the voltages.

C. 1. Using a dc voltmeter, measure the voltages V_p, V_s, V_{D1}, V_{D2}, V_{D3}, V_{D4}, and V_L.

2. Record these values in Table 2-6.

D. 1. Using an ac voltmeter, measure the voltages V_p, V_s, V_{D1}, V_{D2}, V_{D3}, V_{D4}, and V_L.

2. Record these values in Table 2-6.

E. 1. Using the data from step B.1., compute the value of the dc voltage across R_L.

2. Compute the percent difference between the calculated and measured values of the dc voltage across R_L.

F. Why is there a difference between the ac voltages V_s and V_L?

G. Calculate the dc value of current (I_L) through R_L.

TABLE 2-6

Voltages	dc reading in volts	ac reading in volts
V_p		
V_s		
V_{D_1}		
V_{D_2}		
V_{D_3}		
V_{D_4}		
V_L		

H. Compare the peak values of the diode voltage for the two-diode full-wave rectifier circuit and the bridge rectifier circuit. The diodes of which circuit have the largest voltage?

EXTENDED STUDY TOPICS

1. For the half-wave rectifier circuit of figure 2-1, $V_M = 220$ V and R = 5.1 k Ω. Find (a) the dc voltage across R, (b) the minimum PRV rating of the diode, (c) the maximum allowable dc output current of the diode, and (d) the maximum surge current when the circuit is energized.
2. A half-wave rectifier circuit has a dc current of 360 mA through a 1200-ohm load resistance. Find (a) the dc output voltage, (b) the minimum PRV rating of the diode, and (c) the maximum surge current when the circuit is energized.
3. A transformer has a primary voltage of 120 V. The secondary has three voltage windings (250 V, 75 V, 6 V) and a center tap. Find the turns ratio for each voltage winding.
4. If a 2-kΩ load is connected to the 250-V winding of the transformer in problem 3, find (a) I_s, (b) I_p, and (c) P_s and P_p.
5. Repeat problem 4 for the 75-V winding.
6. Repeat problem 4 for the 6-V winding.
7. Repeat problem 4 if the 2-kΩ load is connected from the center tap to one terminal of the 250-V winding.
8. For the full-wave rectifier circuit of figure 2-11, the transformer turns ratio is 20, $V_p = 120$ V, and the dc output current is 15 mA. Find (a) the dc output voltage, (b) the primary transformer current, (c) the reflected impedance, (d) the minimum PRV rating of the diode, and (e) the maximum surge current at turn on.
9. Repeat problem 8 for a bridge rectifier circuit.

Power supply filters

OBJECTIVES

After studying this unit, the student will be able to discuss and demonstrate an understanding of the basic principles of:

- How the RC filter produces a dc voltage level including an ac ripple voltage
- Ripple voltage, ripple factor, and the voltage regulation calculations for RC filters and L section filters
- The use of Schade's curves when designing RC filters
- Voltage doubler circuit operation
- L section filter operation and the critical inductance calculation for this filter

THE RC FILTER

The RC filter consists of a capacitor connected in parallel with a resistor (usually the load). A simplified half-wave rectifier circuit using an RC filter is shown in figure 3-1a, page 42. The diode and source resistance are assumed to be zero.

The voltage waveform which appears across the RC filter is shown in figure 3-1b. The applied voltage, v_{in}, is a sine wave. During the period of one sine wave, the diode allows one positive half-cycle of the sine wave to pass the RC filter. As a result, the capacitor charges to the maximum value of the sine wave, V_M, during the time period T_1. As the sine wave decreases, the capacitor discharges during the time period T_2. The capacitor must discharge through the load resistance R because the diode will not permit the flow of current in the reverse direction. The RC time constant of the filter determines how quickly the capacitor discharges.

If the discharge path for the capacitor is removed from the circuit, that is, if resistor R is removed, then the capacitor will not discharge. In other words, the capacitor will charge to the maximum value of the sine wave, V_M, and will remain at this value. Practically,

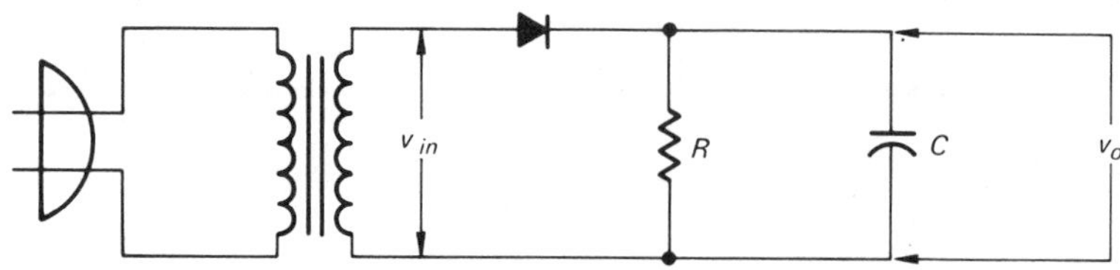

(a) HALF-WAVE RECTIFIER WITH AN RC FILTER

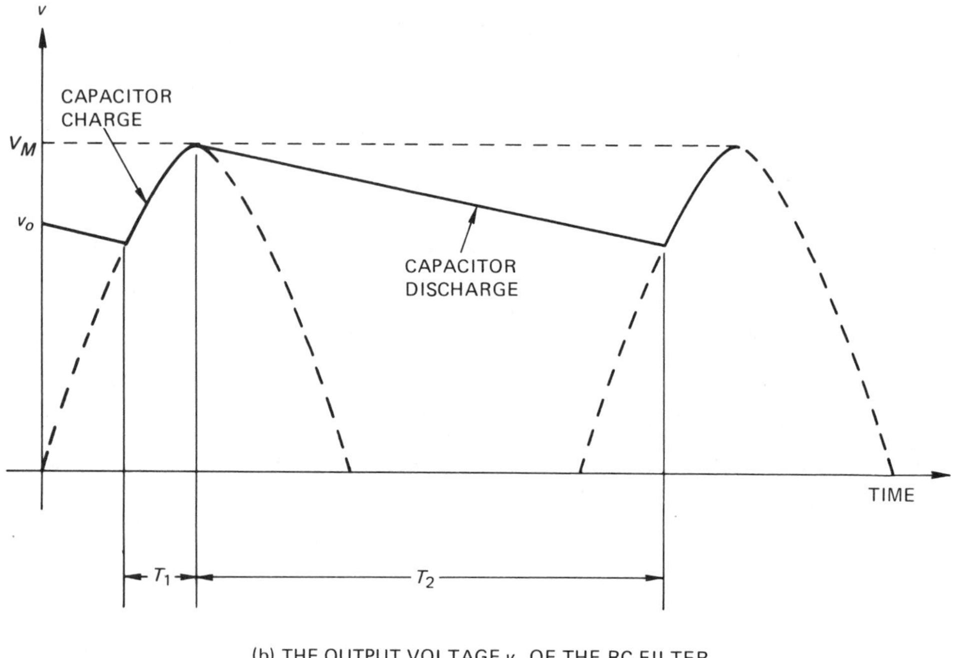

(b) THE OUTPUT VOLTAGE v_o OF THE RC FILTER FOR THE HALF-WAVE RECTIFIER

FIG. 3-1 THE HALF-WAVE RECTIFIER CIRCUIT AND OUTPUT VOLTAGE WAVEFORM v_o

however, a circuit will always contain a load so that the capacitor can discharge.

Figure 3-2a shows a two-diode full-wave rectifier with an RC filter. The diode and source resistances are assumed to be zero. The voltage waveform across the RC filter is shown in figure 3-2b. The applied voltage, v_{in}, is a sine wave. During the period of one sine wave, the diodes allow two positive half-cycles of the sine wave to pass to the RC filter. In a manner similar to that for the half-wave rectifier circuit, the capacitor will charge to the maximum value of the sine wave, V_M, during the time period T_1. As the sine wave decreases, the capacitor discharges during time period T_2. The capacitor must discharge through the load resistance R. The RC time constant of the filter determines how quickly the capacitor discharges.

When the capacitor is charging,the diode conducts for a period of time (T_1) which is short compared to the capacitor discharge time (T_2). While discharging, the capacitor supplies the load with the average or dc current applied to it during charging. In other words, the average (or dc) current supplied to the capacitor during the charge period must equal the average (or dc) current during the discharge

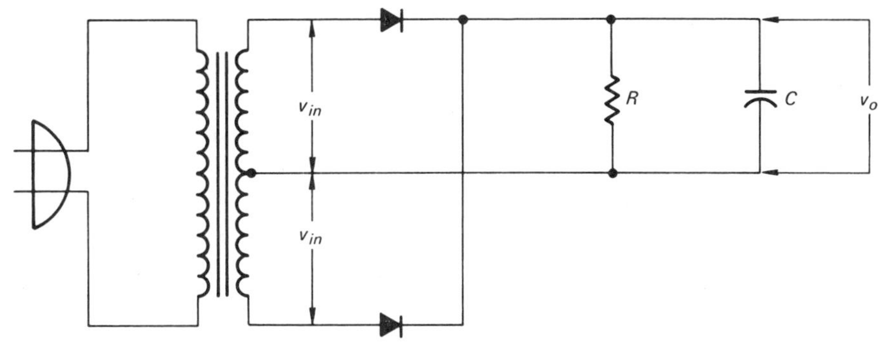

(a) FULL-WAVE RECTIFIER WITH RC FILTER

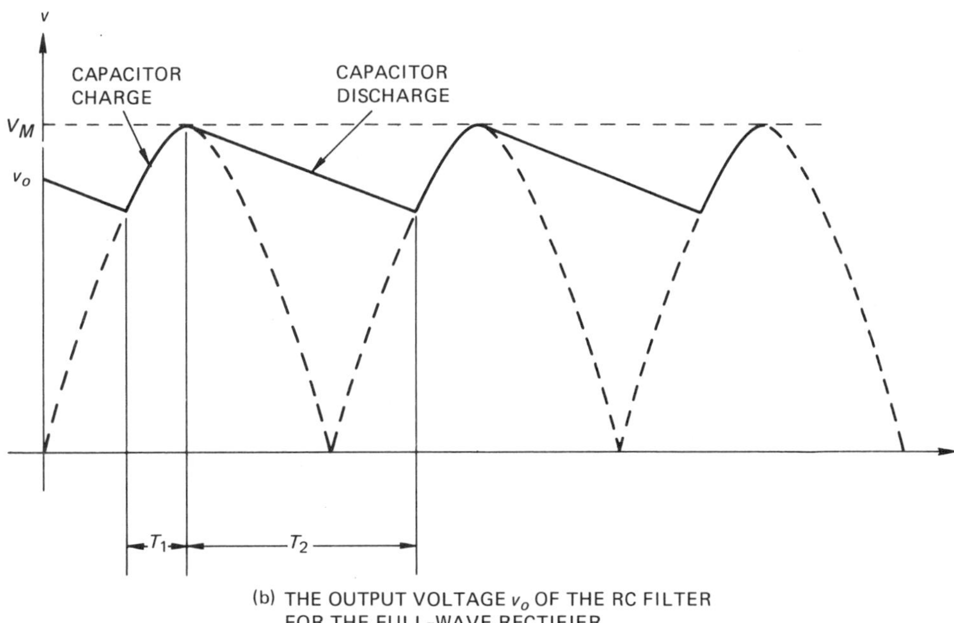

(b) THE OUTPUT VOLTAGE v_o OF THE RC FILTER FOR THE FULL-WAVE RECTIFIER

FIG. 3-2 THE FULL-WAVE RECTIFIER CIRCUIT AND OUTPUT VOLTAGE WAVEFORM v_o

period. To achieve this dc current value, the current peaks during diode conduction must be very much larger than the current peaks during the capacitor discharge period. Another way of stating the above condition is that there must be larger peak currents in the smaller time T_1 as compared to the peak currents in time T_2. Figure 3-3, page 44, shows the diode current waveform and the output voltage waveform for a half-wave rectifier.

RIPPLE VOLTAGE, RIPPLE FACTOR, AND VOLTAGE REGULATION

The RC filter output voltage has a dc voltage level and an ac variation called the *ripple* voltage, figure 3-4, page 44. The ripple voltage arises from the capacitor charge and discharge actions. Similar to any ac voltage, the ripple voltage has a peak-to-peak value. To measure the dc voltage level, a dc voltmeter can be connected across the RC filter

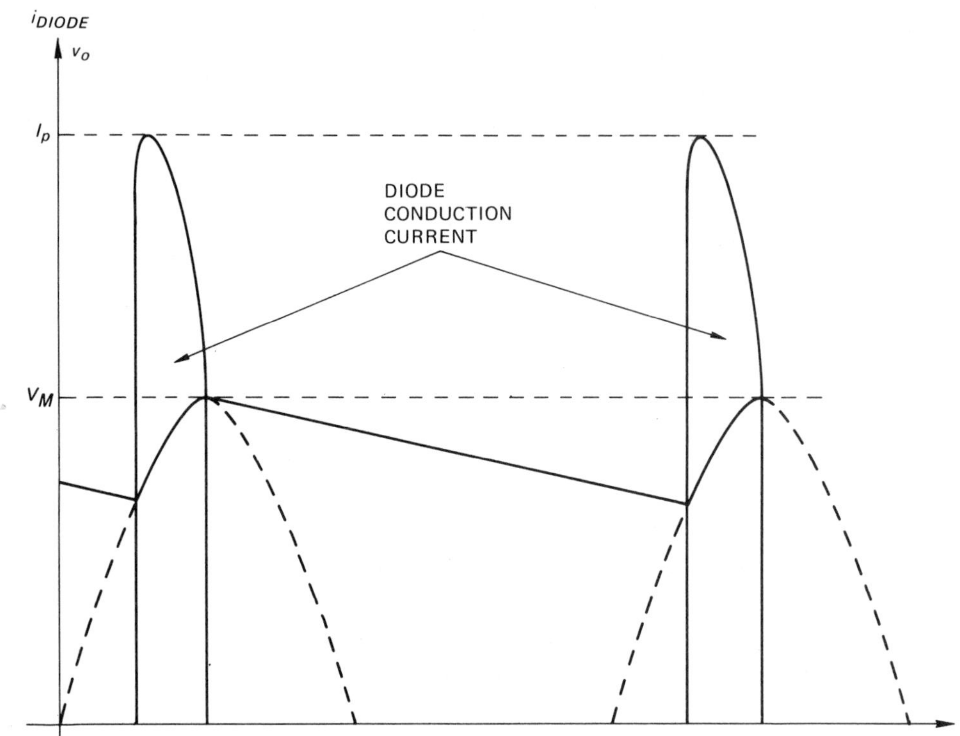

FIG. 3-3 DIODE CONDUCTION CURRENT FOR A HALF-WAVE RECTIFIER

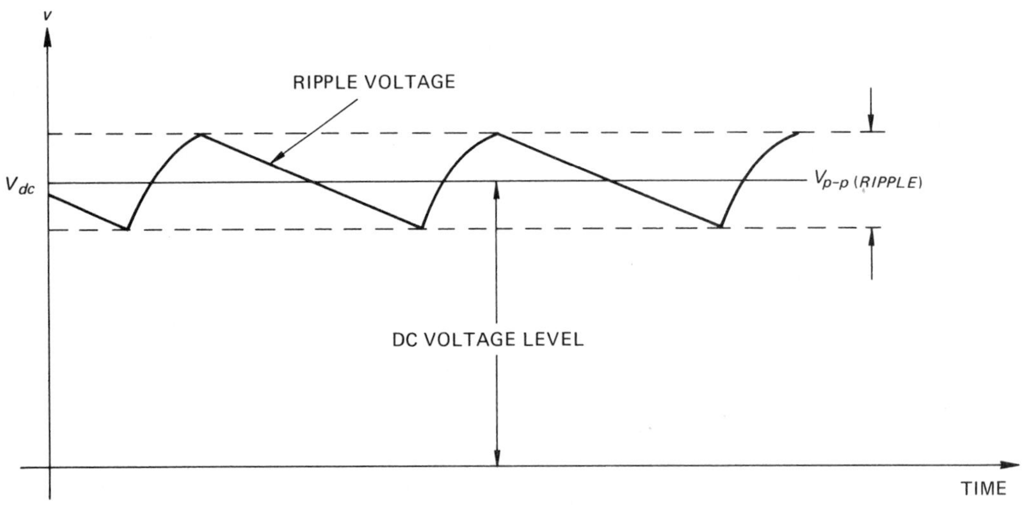

FIG. 3-4 THE RIPPLE VOLTAGE OF AN RC FILTER

output. An ac (rms) meter connected across the RC filter will indicate the rms value of the ripple voltage. That is, the ac (rms) meter will block out the dc voltage level so that it indicates the effect of the ripple voltage only.

Any discussion of power supply filters must include a quantity called the *percent ripple factor*, % γ (The symbol γ is the Greek letter gamma). Eq. 3.1 defines the quantity % γ.

$$\% \gamma = \frac{v_r \text{ (rms)}}{V_{dc}} \times 100 \qquad \text{Eq. 3.1}$$

where $v_{r\,(rms)}$ = the rms value of the ripple voltage (measured by an ac voltmeter which blocks any dc voltage).

V_{dc} = the dc output voltage level (measured by a dc voltmeter).

$\%\gamma$ = the percent ripple factor. This quantity compares the ripple voltage to the dc voltage level.

PROBLEM 1

The ripple voltage in figure 3-4 is a sine wave and has a peak-to-peak value of 4 V. The dc voltage level is 50 V. Find a. $V_{r\,(rms)}$, and b. $\%\gamma$.

a. $V_{r\,(rms)}$ = 0.5 $V_{p\text{-}p}$ x 0.707 (for a sine wave)

= 0.5 x 4 V x 0.707

= **1.414 V**

b. $\%\gamma = \dfrac{V_{r\,(rms)}}{V_{dc}} \times 100$

$= \dfrac{1.414\ \text{V} \times 100}{50\ \text{V}}$

= **2.828** %

If the ripple voltage resembles a sine wave, then Problem 1 shows the typical calculations requred to find the percent ripple factor. The ripple voltage waveform usually resembles either a sine wave or a sawtooth (triangular) waveshape, figure 3-5. When the ripple voltage has a sawtooth waveshape, the ripple factor depends upon several parameters, including the frequency, capacitance, and resistance of the RC filter. The mathematical analysis of the ripple factor for the sawtooth waveshape is beyond the scope of this text, but can be found in reference no. 42 of the Bibliography in the Instructor's Guide. Eq. 3.2 and Eq. 3.3 define the percent ripple factor for half-wave and full-wave rectifiers when the ripple voltage is a sawtooth. For a half-wave rectifier,

$$\%\gamma\,(\text{St-hw}) = \frac{100}{2\sqrt{3}\ f_{in}\ C\,R} \qquad \textbf{Eq. 3.2}$$

where f_{in} = input frequency of the sine wave, in hertz

C = filter capacitance, in farads

R = load resistance, in ohms

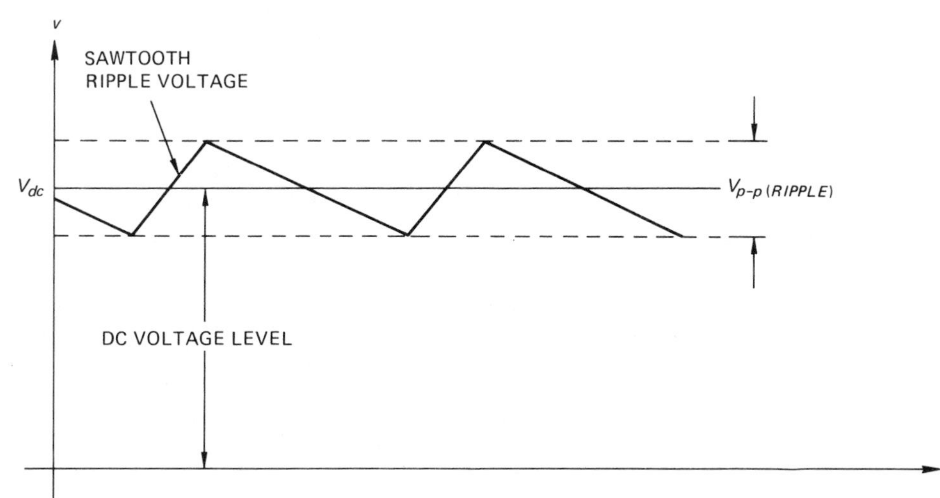

FIG. 3-5 THE RIPPLE VOLTAGE AS A SAWTOOTH WAVESHAPE

$\% \gamma$ (St-hw) = the percent ripple factor for a half-wave rectifier RC filter when the output voltage resembles a sawtooth waveshape

For a full-wave rectifier,

$$\% \gamma \text{(St-fw)} = \frac{100}{4\sqrt{3}\ f_{in}\ C\ R} \qquad \text{Eq. 3.3}$$

A comparison of Eq. 3.2 and Eq. 3.3 shows that the full-wave rectifier circuit has a value for the percent ripple factor equal to half that of the half-wave rectifier circuit. This factor represents another advantage of the full-wave rectifier circuit over the half-wave rectifier circuit.

PROBLEM 2

A full-wave rectifier circuit and a half-wave rectifier circuit each have an input frequency of 60 Hz, a load resistance of 5000 ohms, and a capacitor with a value of 19.3 μ F. Assuming a sawtooth waveshape for the ripple voltage, find the $\% \gamma$ for each circuit.

Half-wave rectifier:

$$\% \gamma \text{(St-hw)} = \frac{100}{2\sqrt{3} \times 60 \times 19.3 \times 10^{-6} \times 5000} = 5\%$$

Full-wave rectifier:

$$\% \gamma \text{(St-fw)} = \frac{\% \gamma \text{(St-hw)}}{2} = 2.5\%$$

A second quantity of interest in any discussion of power supply filters is the amount of change in the output dc voltage over the range of the circuit operation. In other words, how much does the dc voltage change across the output of a power supply filter when going from a no-load condition to a full-load condition. This voltage change is called *voltage regulation* and is defined as follows:

$$\% \text{V.R.} = \frac{V_{NL} - V_{FL}}{V_{FL}} \times 100 \qquad \text{Eq. 3.4}$$

where V_{NL} = voltage at no load (open circuit, no current flow)

V_{FL} = voltage at full load (a voltage for a given current flow)

%V.R. = percent voltage regulation

PROBLEM 3

A dc power supply has an open circuit voltage of 100 V. When the full-load current is drawn, the output drops to 80 V. Find % V.R.

$$\% \text{V.R.} = \frac{V_{NL} - V_{FL}}{V_{FL}} \times 100$$

$$= \frac{100\text{ V} - 80\text{ V}}{80\text{ V}} \times 100$$

$$= \frac{2000}{80}\% = 25\%$$

Assume the ripple voltage in figure 3-4 is a sine wave with a peak-to-peak value of 0.1 V, and a dc level of 10 V. Find the % γ. (R3-1)

Given a power supply filter circuit, what measurements must be made to determine % γ? (R3-2)

RC FILTER DESIGN PROBLEMS

The RC filter presentation in this text has been simplified to assist the student in understanding the physical phenomenon of RC filter circuits. The mathematical and graphical analysis of RC filters is very complex and beyond the scope of this text. Reference 43 in the Bibliography in the Instructor's Guide contains the derivation of the equation which was solved graphically by O.H. Schade. In 1943, Schade published the graphical results of his work in the Proceedings of the Institute of Radio Engineers (IRE), volume 31. Several of the Schade curves are shown in figures 3-6, 3-7, 3-8, and 3-9 and will be used in solving the problems in this section of the unit.

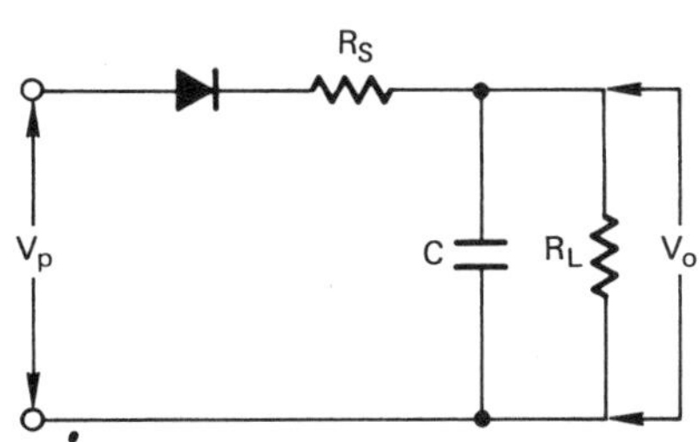

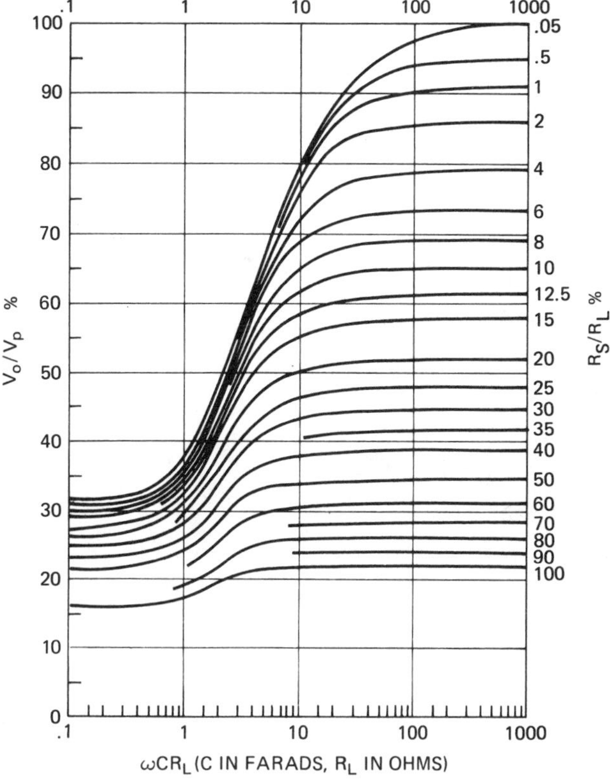

FIG. 3-6 RELATION OF APPLIED ALTERNATING PEAK VOLTAGE TO DIRECT OUTPUT VOLTAGE IN HALF-WAVE CAPACITOR-INPUT CIRCUITS.

(Courtesy Institute of Electrical and Electronics Engineers, Inc.)

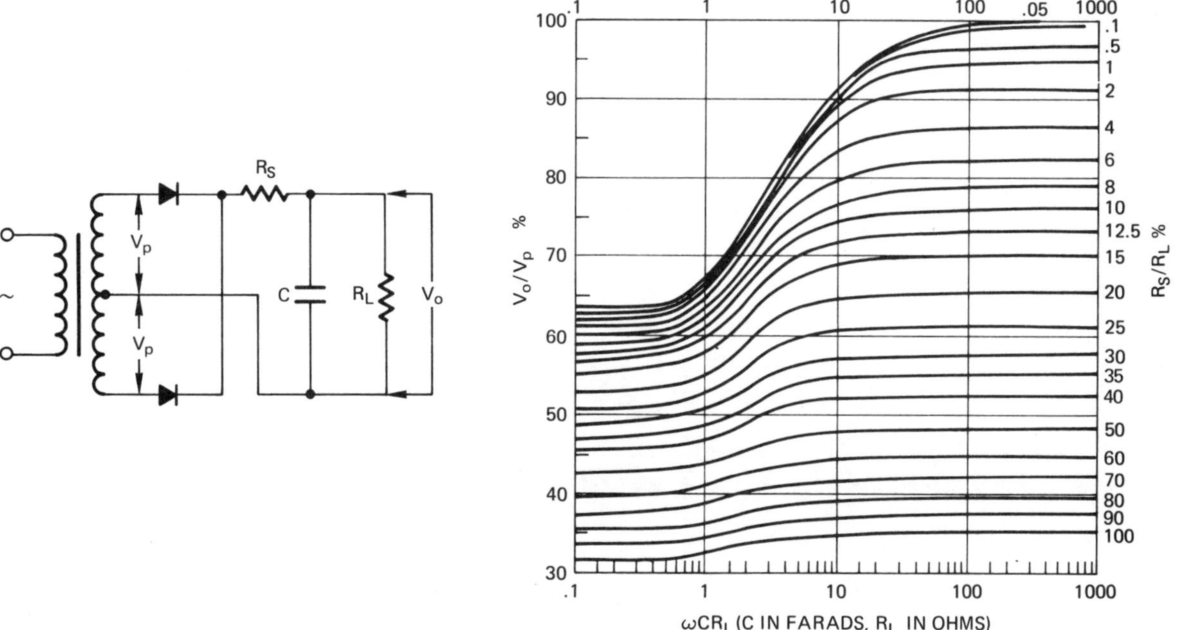

FIG. 3-7 RELATION OF APPLIED ALTERNATING PEAK VOLTAGE TO DIRECT OUTPUT VOLTAGE IN FULL-WAVE CAPACITOR-INPUT CIRCUITS.

(Courtesy Institute of Electrical and Electronics Engineers, Inc.)

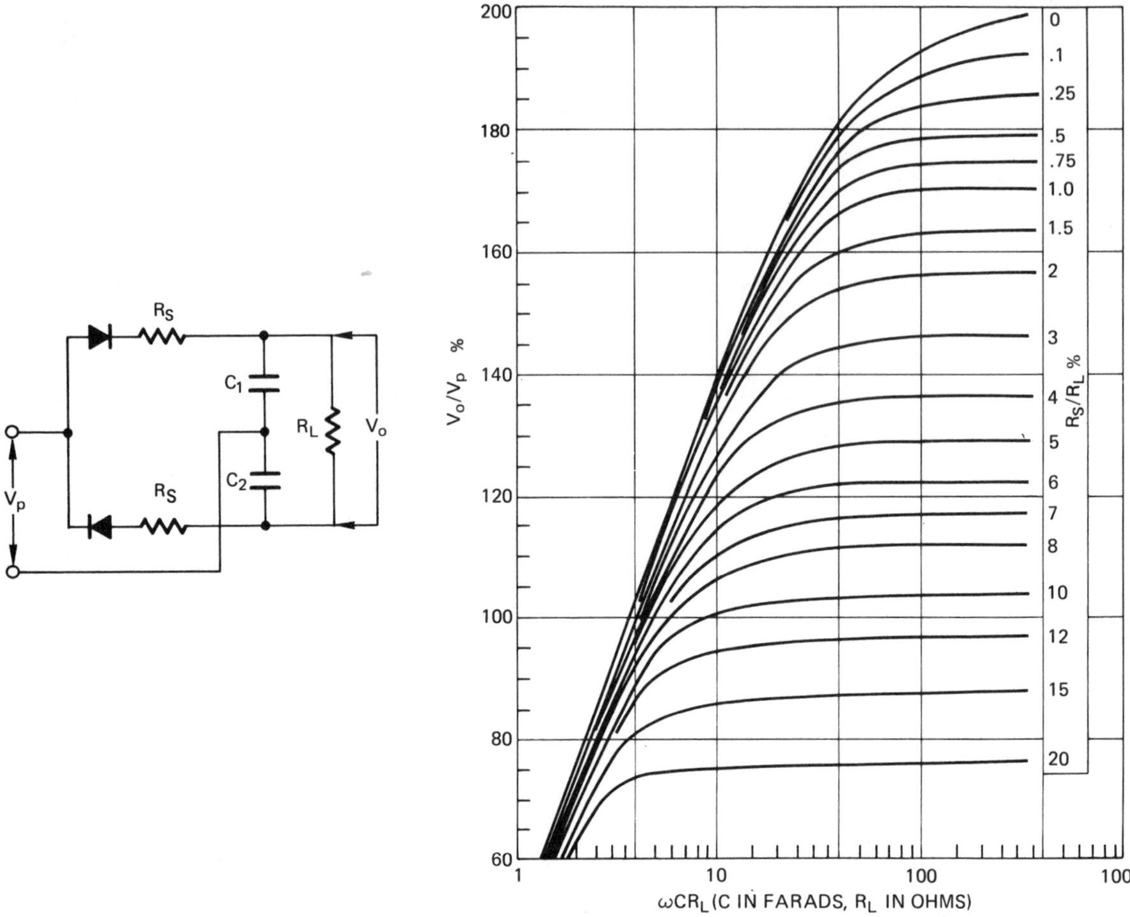

FIG. 3-8 RELATION OF APPLIED ALTERNATING PEAK VOLTAGE TO DIRECT OUTPUT VOLTAGE IN CAPACITOR-INPUT VOLTAGE DOUBLER CIRCUITS.

(Courtesy Institute of Electrical and Electronics Engineers, Inc.)

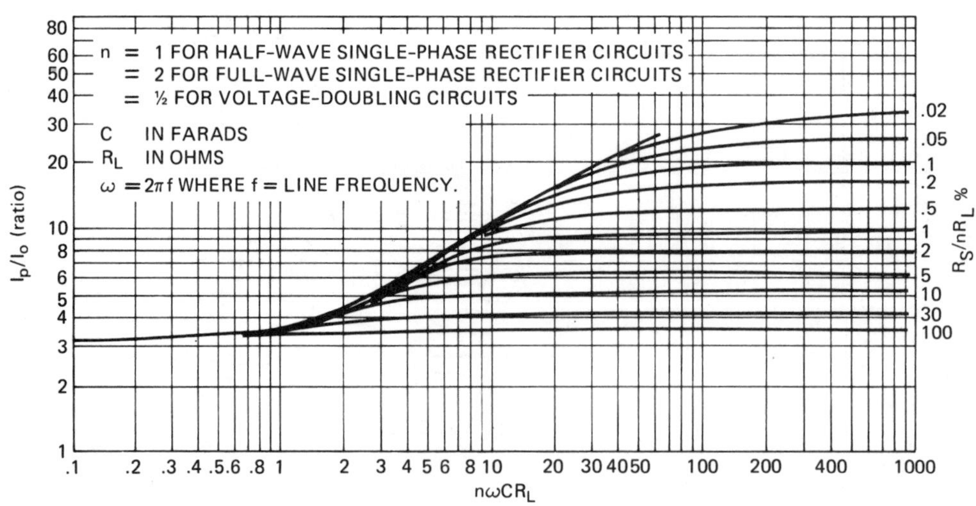

FIG. 3-9 RELATION OF PEAK CURRENT TO AVERAGE CURRENT PER RECTIFIER IN CAPACITOR-INPUT CIRCUITS.

(Courtesy Institute of Electrical and Electronics Engineers, Inc.)

The parameters indicated on the Schade curves are shown in figure 3-10 in terms of measurable or observable physical quantities for the full-wave rectifier circuit containing an RC filter. The independent variable for each of the Schade curves is the dimensionless quantity ωCR_L. The quantity $\omega = 2\pi f$ is usually constant for a given circuit. The load R_L may not appear as a single resistor in the circuit, but may represent the relation of $V_o/I_L = R_L$. If either V_o or I_L changes, R_L must change, therefore ωCR_L will change.

PROBLEM 4.

For the circuit shown in figure 3-10, f_{in} = 60 Hz, R_s = 1 kΩ, I_o = 2.5 mA, V_o = 100 V, and C = 10 μ F. Find: a. R_L, b. R_s/R_L, c. ωCR_L, d. the peak and rms secondary voltage, e. PRV, f. the surge current I_s at turn on, and g. the peak diode current I_p.

a. $$R_L = \frac{V_o}{I_L} = \frac{V_o}{n\,I_o} = \frac{100\text{ V}}{2 \times 2.5\text{ mA}}$$

$$= \mathbf{20\ k\Omega}$$

b. $$\frac{R_s}{R_L} = \frac{1\text{ k}\Omega}{20\text{ k}\Omega} = 0.05 = \mathbf{5\%}$$

c. $$\omega CR_L = 2\pi f_{in} CR_L = 2\pi \times 60 \times 10 \times 10^{-6} \times 20 \times 10^3$$
$$= 2\pi \times 60 \times 20 \times 10^{-2}$$
$$= 24\pi = \mathbf{75}$$

d. From figure 3-7, at R_s/R_L = 5% and ωCR_L = 75, the V_o/V_p curve gives a value of 84%. Thus, V_o = 0.84 x V_p and $V_p = \frac{V_o}{0.84} = \frac{100\text{ V}}{0.84}$

V_p = **119 V (the peak or maximum secondary voltage)**

V_{rms} = 0.707 x V_p = 0.707 x 119 V = **84.1 V**

e. PRV = 2 V_p = **238 V.** The diodes must be rated at this minimum voltage in the reverse voltage direction.

f. $$I_s = \frac{V_p}{R_s} = \frac{119\text{ V}}{1\text{ k}\Omega} = \mathbf{119\ mA}$$

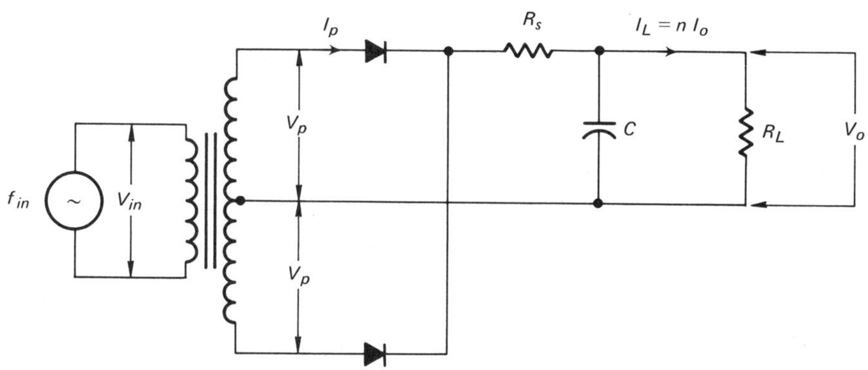

V_p = THE PEAK OR MAXIMUM SECONDARY SINE-WAVE VOLTAGE
I_L = THE DC OUTPUT OR LOAD CURRENT THROUGH R_L
V_o = THE DC OUTPUT OR LOAD VOLTAGE ACROSS R_L
I_L = $n\,I_o$
I_o = THE AVERAGE OR DC DIODE RECTIFIER FORWARD CURRENT
n = 1.0 FOR A HALF-WAVE RECTIFIER
= 2.0 FOR A FULL-WAVE RECTIFIER
= 0.5 FOR A VOLTAGE DOUBLER CIRCUIT
I_p = THE PEAK DIODE CONDUCTION CURRENT PER DIODE RECTIFIER
R_s = A RESISTANCE WHICH INCLUDES THE DIODE FORWARD RESISTANCE, THE TRANSFORMER WINDING RESISTANCE, AND ANY OTHER RESISTANCE IN SERIES WITH THE DIODES.

FIG. 3-10 FULL-WAVE RECTIFIER CIRCUIT WITH AN RC FILTER

I_S is the maximum current the diodes experience when the circuit is turned on.

g. The peak current I_P to the diodes is determined through the use of the Schade curve in figure 3-9.

n = 2 for a full-wave rectifier

R_S/nR_L = 5%/2 = 2.5%

$n\omega CR_L$ = 2 x 75 = 150

I_P/I_O = 7.9 (from figure 3-9)

I_P = 7.9 x I_O = 7.9 x 2.5 mA = **19.75 mA**

Repeat Problem 4 for a half-wave rectifier circuit. (R3-3)

Repeat Problem 4 for a full-wave rectifier circuit using the same parameters except $C = 0.1\ \mu F$, and I_O = 4 mA. (R3-4)

Repeat Problem 4 for a half-wave rectifier circuit if $C = 1.0\ \mu F$, and I_O = 4 mA. (R3-5)

VOLTAGE DOUBLER CIRCUIT

A full-wave voltage doubler circuit is shown in figure 3-11a. During the positive half-cycle of the input sine wave, v_{in}, diode D_1 conducts and capacitor C_1 is charged to a peak voltage V_M, figure 3-11b. For the same positive half cycle, diode D_2 is reverse biased and is not conducting.

Diode D_2 conducts during the negative half-cycle of the input sine wave, v_{in}, and capacitor C_2 is charged to a peak voltage V_M as shown in figure 3-11c. For the same negative half-cycle, diode D_1 is reverse biased and is not conducting.

In the no-load condition, the output voltage across the two capacitors is 2 V_M. If a load is connected to the output of the voltage doubler, the voltage across the two capacitors decreases.

The effective capacitance of C_1 and C_2 in series is less than the capacitance of either C_1 or C_2 alone. The lower effective capacitance results in a poorer filtering action than that provided by the single capacitor in a regular full-wave rectifier circuit. For a full-wave voltage doubler circuit, the PRV across each diode is 2 V_M.

PROBLEM 5.

For the voltage doubler circuit shown in figure 3-12, f_{in} = 60 Hz, R_S = 1 kΩ, I_O = 2.5 mA, V_O = 100 V, and $C_1 = C_2 = 10\ \mu F$. Find: a. R_L, b. R_S/R_L, c. ωCR_L, d. the peak and rms secondary voltage, e. PRV, and f. the peak diode current I_P.

a. $R_L = \dfrac{V_O}{I_L} = \dfrac{V_O}{nI_O} = \dfrac{100\text{ V}}{0.5 \times 2.5\text{ mA}}$ = **80 kΩ**

b. $\dfrac{R_S}{R_L} = \dfrac{1\text{ k}\Omega}{80\text{ k}\Omega}$ = 0.0125 = **1.25%**

$\omega CR_L = 2\pi f_{in} CR_L$

C = C_1 in series with C_2

$C = \dfrac{10\ \mu F}{2} = 5\ \mu F$

$\omega CR_L = 2\pi \times 60 \times 5 \times 10^{-6} \times 80 \times 10^3$

= **150**

d. From figure 3-8, at R_S/R_L = 1.25% and ωCR_L = 150, the value of V_O/V_P is 168%. Thus, $V_O = 1.68 \times V_P$ or

$V_P = \dfrac{V_O}{1.68} = \dfrac{100\text{ V}}{1.68}$

V_P = **59.5 V**

V_{rms} = 0.707 x V_P = 0.707 x 59.5
= **42.1 V**

e. PRV = 2 V_P = 2 x 59.5 V = **119 V**
The PRV is the minimum voltage rating of the diodes in the reverse voltage direction.

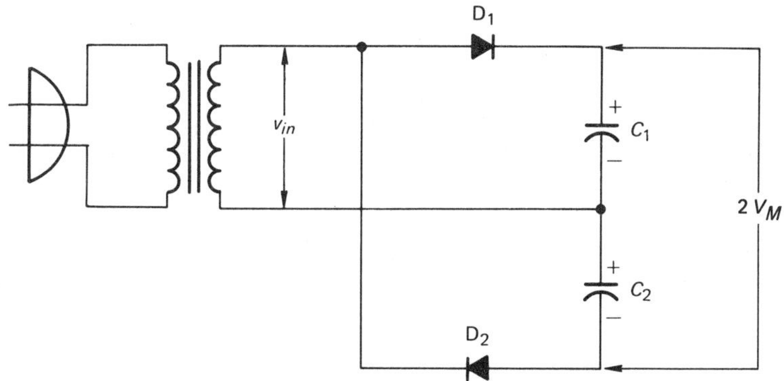

(a) VOLTAGE DOUBLER CIRCUIT

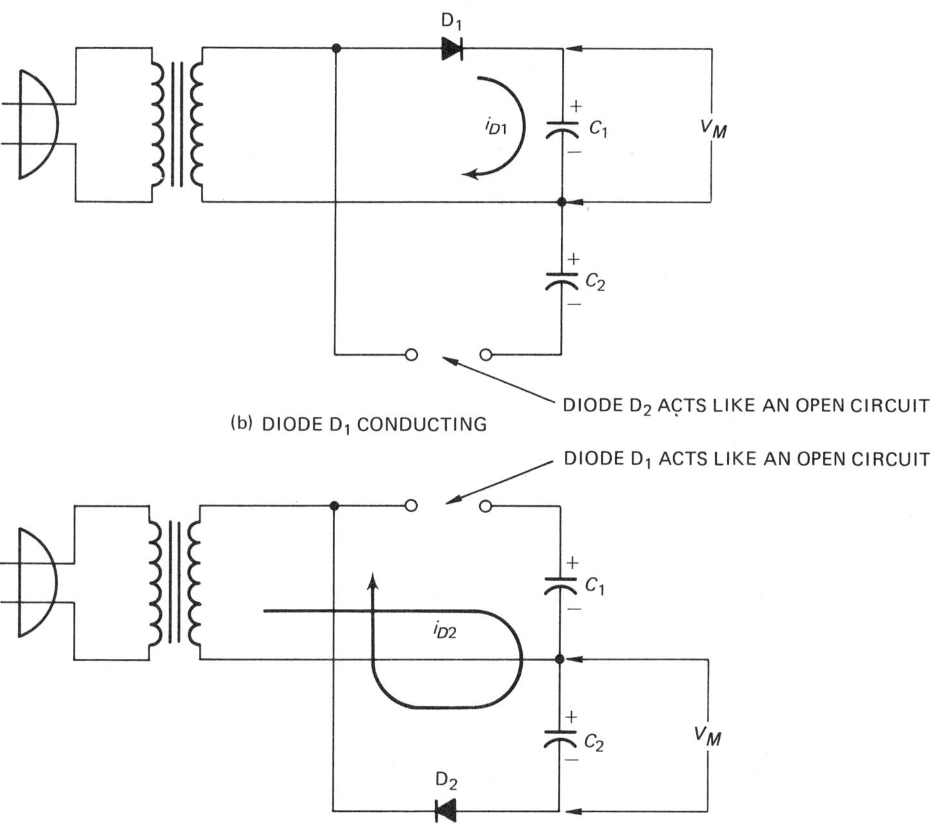

(c) DIODE D_2 CONDUCTING

FIG. 3-11 FULL-WAVE VOLTAGE DOUBLER CIRCUIT

f. The peak current, I_p, experienced by the diodes is determined from the use of the Schade curves in figure 3-9.

$n = 0.5$ for a voltage doubler

$R_s/n\,R_L = 1.25\,\%/0.5 = 2.5\%$

$n\,\omega\,C\,R_L = 0.5 \times 150 = 75$

$I_p/I_o = 7.9$ (from figure 3-9)

$I_p = 7.9 \times I_o = 7.9 \times 2.5$ mA

$= \mathbf{19.75\ mA}$

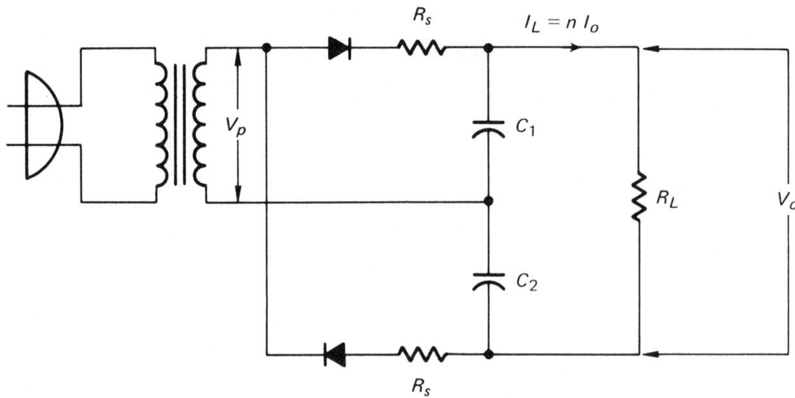

FIG. 3-12 VOLTAGE DOUBLER CIRCUIT

Repeat Problem 5 with I_O = 15 mA, and V_O = 75 V. (R3-6)

Repeat Problem 5 with $C_1 = C_2 = 1\ \mu F$. (R3-7)

CHOKE INPUT OR L SECTION FILTER

The L section filter shown in figure 3-13 provides low ripple voltage at large load currents. An advantage of the L section filter over the RC filter is the presence of the inductor in series with the diode. In this arrangement, the diode can conduct continuously.

The dc voltage developed by the L section filter across R_L is 0.318 V_M for a half-wave rectifier and 0.636 V_M for a full-wave rectifier. The inductor in the circuit of figure 3-13 offers a high impedance to ac signals. In addition, the capacitor offers a low parallel impedance to ac signals. In other words, at point 1 in figure 3-13, both ac and dc signals are present. The inductor passes dc voltage easily and opposes ac voltage. If an ac voltage reaches point 2 with the dc voltage, then the capacitor presents a short circuit to the ac back to the source. As a result, the filtering action for ac voltage of the L section filter is greater than that of either the capacitor or inductor alone.

Since the inductor helps to filter (prohibit or choke) more ac voltage than the capacitor alone, the inductor is also known as a *choke* and the filter circuit is called a choke input filter.

To understand the function of the L section filter in a circuit, the dc and ac operation of the filter must be analyzed.

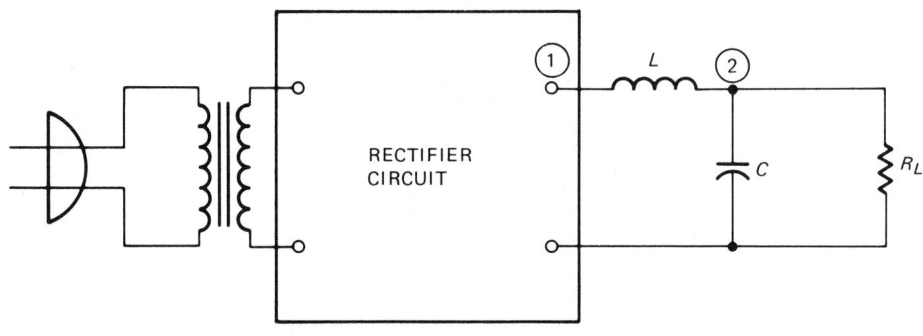

FIG. 3-13 AN L SECTION FILTER CIRCUIT

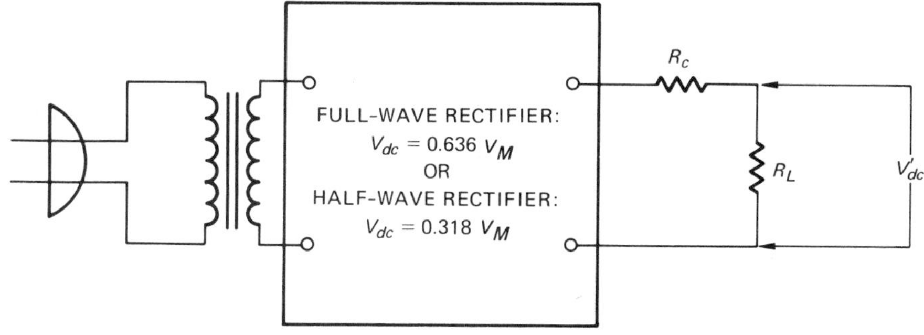

(a) THE DC EQUIVALENT CIRCUIT OF AN L SECTION FILTER

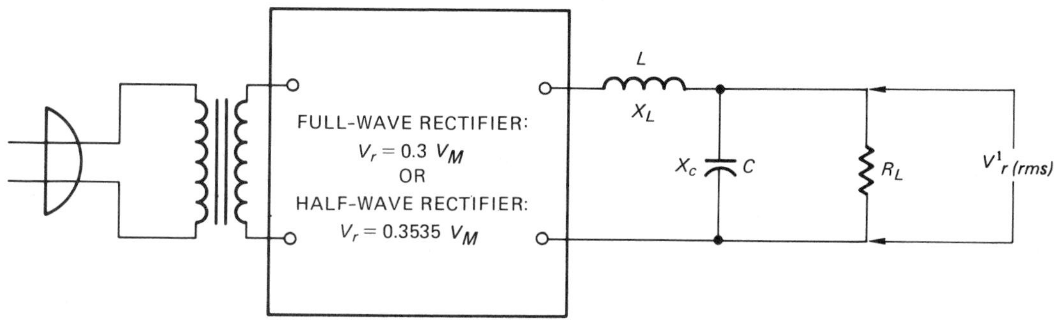

(b) THE AC EQUIVALENT CIRCUIT OF AN L SECTION FILTER

FIG. 3-14

Figure 3-14a shows the dc equivalent circuit of the L section filter connected to a load resistor R_L. The resistor R_c is the dc resistance of the choke (inductor L). The output dc voltage can be calculated in terms of the dc voltage drops in the circuit.

$$V'_{dc} = 0.318\ V_M - I_{dc}\ R_c \quad \text{(half-wave rectifier)} \qquad \text{Eq. 3.5a}$$

$$V'_{dc} = 0.636\ V_M - I_{dc}\ R_c \quad \text{(full-wave rectifier)} \qquad \text{Eq. 3.5b}$$

PROBLEM 6.

The dc current through R_L in an L section filter is 10 mA and the maximum sine wave voltage is 314 V. The dc resistance of the choke is 500 Ω. Find the dc output voltage for a half-wave rectifier and a full-wave rectifier.

Half-wave rectifier:

$$V'_{dc} = 0.318\ V_M - I_{dc}\ R_c$$
$$= 0.318 \times 314\ \text{V} - 0.01\ \text{A} \times 500\ \Omega$$
$$= 100\ \text{V} - 5\ \text{V} = \mathbf{95\ V}$$

Full-wave rectifier:

$$V'_{dc} = 0.636\ V_M - I_{dc}\ R_c$$
$$= 0.636 \times 314\ \text{V} - 0.01\ \text{A} \times 500\Omega$$
$$= 200\ \text{V} - 5\ \text{V} = \mathbf{195\ V}$$

The ac operation of the L section filter circuit is determined by the use of the ac equivalent circuit shown in figure 3-14b. The value of the resistor R_L is considered to be much larger than X_c; similarly, X_L is much larger than X_c. Eq. 3.6 for $V_{r(rms)}$ is obtained using the voltage divider rule for the ac equivalent circuit and the above approximations.

$$V_{r\,(rms)} = \frac{X_c}{X_L}\ V_{r\,(rms)} \qquad \text{Eq. 3.6}$$

The voltage $V_{r(rms)}$ in Eq. 3.6 is obtained by a mathematical technique called

Fourier analysis which is beyond the scope of this text. The result of the Fourier analysis is given in Eq. 3.7a and Eq. 3.7b.

$$V_{r\,(rms)} = 0.3\ V_M \quad \text{(full-wave rectifier)} \qquad \text{Eq. 3.7a}$$

$$V_{r\,(rms)} = 0.3535\ V_M \quad \text{(half-wave rectifier)} \qquad \text{Eq. 3.7b}$$

If Eq. 3.7a is substituted into Eq. 3.6 with $X_C = 1/(2\pi f_{out}C)$ and $X_L = 2\pi f_{out}L$, where $f_{out} = 2f_{in}$ for a full-wave rectifier, then the following equation results:

Full-wave rectifier:

$$V_{r\,(rms)} = \frac{1901}{f^2_{in}\,L\,C}\ V_M \qquad \text{Eq. 3.8}$$

where

- f_{in} = input frequency into the rectifier circuit, in hertz
- L = inductance, in henries
- C = capacitance, in microfarads
- V_M = maximum value of the input sine wave
- $V'_{r\,(rms)}$ = the rms value of the ripple voltage across R_L for an L section filter, in volts.

If Eq. 3.7b is substituted into Eq. 3.6 with $X_C = 1/(2\pi f_{out}C)$, and $X_L = 2\pi f_{out}L$, where $f_{out} = f_{in}$ for a half-wave rectifier, then the following equation results:

Half-wave rectifier:

$$V_{r\,(rms)} = \frac{8964}{f^2_{in}\,L\,C}\ V_M \qquad \text{Eq. 3.9}$$

where

- f_{in} = input frequency into the rectifier circuit, in hertz
- L = inductance, in henries
- C = capacitance, in microfarads
- V_M = maximum value of the input sine wave
- $V'_{r\,(rms)}$ = the rms value of the ripple voltage across R_L for an L section filter, in volts.

To calculate the output ripple factor for the L section filter with a half-wave rectifier, first substitute Eq. 3.9 for the numerator of Eq. 3.1. Substitute for the denominator of Eq. 3.1 the quantity $V'_{dc} = 0.318\ V_M$. The resulting equation is:

Half-wave:

$$\%\gamma' = \frac{2\ 818\ 900}{f^2_{in}\,L\,C} \qquad \text{Eq. 3.10a}$$

The parameters f_{in}, L, and C are defined for Eq. 3.9. If f_{in} = 60 Hz, then Eq. 3.10a becomes:

Half-wave when f_{in} = 60 Hz:

$$\%\gamma' = \frac{785}{L\,C} \qquad \text{Eq. 3.10b}$$

To calculate the output ripple factor for the L section filter with a full-wave rectifier, first substitute Eq. 3.8 for the numerator of Eq. 3.1. Substitute for the denominator of Eq. 3.1 the quantity $V'_{dc} = 0.636\ V_M$. The resulting equation is:

Full-wave:

$$\%\gamma' = \frac{298\ 900}{f^2_{in}\,L\,C} \qquad \text{Eq. 3.10c}$$

The parameters f_{in}, L, and C are defined for Eq. 3.9. If f_{in} = 60 Hz, then Eq. 3.10c becomes:

Full-wave when f_{in} = 60 Hz:

$$\%\gamma' = \frac{83}{L\,C} \qquad \text{Eq. 3.10d}$$

PROBLEM 7.

An L section filter is connected to a full-wave rectifier input. The sine wave has a frequency of 60 Hz and an rms voltage of 120 V. The filter parameters are L = 10 H, C = 50 μF, and R_L = 5 kΩ. Find: a. the dc output voltage if R_c = 0, b. the rms ripple voltage, and c. %γ.

a. $V'_{dc} = 0.636\ V_M - I_{dc} R_c$

$= 0.636\ (1.414 \times 120\ V) - 0$

$= \mathbf{108\ V}$

b. $V_{r\,(rms)} = \dfrac{1901\ (V_M)}{f^2_{in}\ L\ C}$

$= \dfrac{1901\ (1.414 \times 120\ V)}{(60)^2 \times 10 \times 50}$

$= \mathbf{0.161\ V}$

c. $\% \gamma = \dfrac{83}{10 \times 50} = \mathbf{0.166\%}$

An L section filter is connected to a half-wave rectifier. If the maximum rectifier voltage V_M is 200 V, the choke resistance is 100 Ω, and the load current is 75 mA, find the dc output voltage of the filter. *(R3-8)*

Repeat (R3-8) for a full-wave rectifier. (R3-9)

CRITICAL INDUCTANCE FOR L FILTERS

A minimum inductance value must be maintained when using L section filters with a full-wave rectifier to keep the diodes conducting at all times. This minimum inductance value is called *critical inductance.* If there is no inductor in the circuit, the result is an RC filter with large peak currents charging the capacitor. As inductance is added to the circuit, the large peak currents decrease until eventually the diode supplies current continuously and no cutoff occurs.

If the diode is to conduct continuously, the maximum current value of the ac component, I_M, must not exceed the dc current, as shown in figure 3-15b, page 56. Figure 3-15a shows the resulting diode conduction when the current values are less than the critical inductance. The minimum value of the critical inductance can be found by equating the dc current value, I_{dc}, to the ac component, I_M, for a full-wave rectifier input.

The derivation of the critical inductance equations again involves Fourier analysis. The results of the analysis are shown below.

Full-wave rectifier:

$$L_C \geqq \frac{R_L}{6 \pi f_{in}} \qquad \textbf{Eq. 3.11}$$

where

R_L = load resistance of the L section filter, in ohms

f_{in} = input sine-wave frequency into the rectifier circuit, in hertz

L_C = critical inductance for an L filter with a full-wave rectifier, in henries

The equations resulting from the Fourier analysis show that the value of critical inductance depends on the load connected to the power supply at any one operating frequency. As the load current increases (smaller R_L), the value of the critical inductance decreases. This action is reasonable since a larger inductor must be used to limit the ac current to a smaller swing for smaller values of the dc load current. This condition insures that the diode will always conduct.

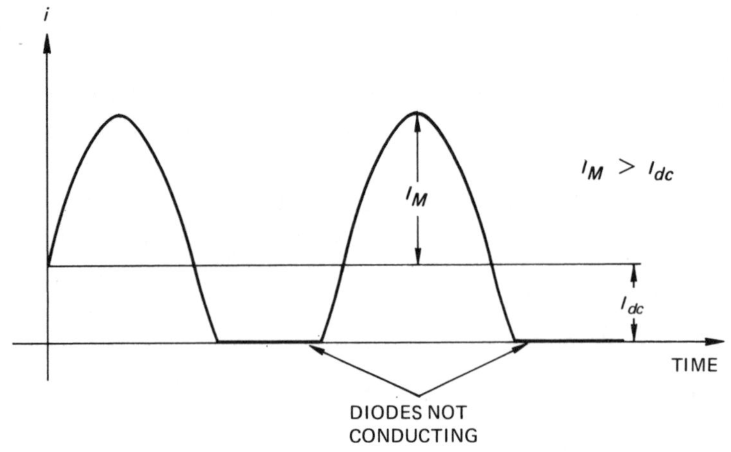

(a) BELOW CRITICAL L VALUE

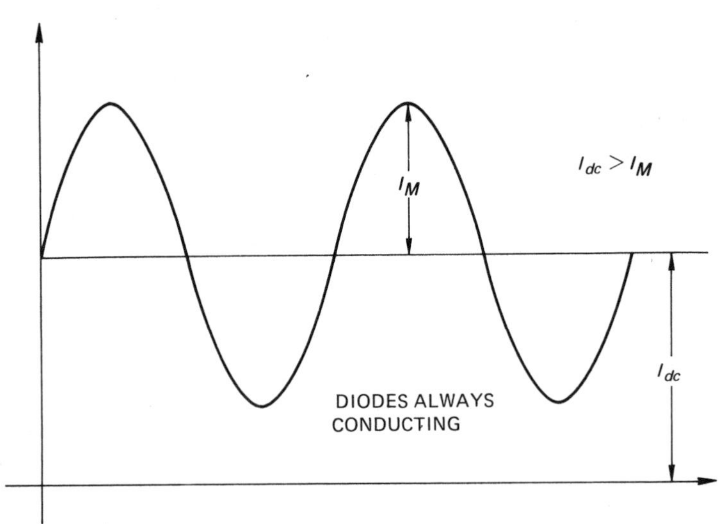

(b) ABOVE CRITICAL L VALUE

FIG. 3-15 DIODE CONDUCTION FOR CRITICAL *L*

PROBLEM 8.

An L section filter must provide a dc current of 120 mA at 36 V with a percent ripple of less than 2%. The sine-wave frequency into the rectifier is 60 hertz. Design an L section filter to meet the given requirements for a full-wave rectifier.

a. $R_L = \dfrac{V'_{dc}}{I_{dc}} = \dfrac{36\text{ V}}{120\text{mA}} = \mathbf{0.300\ k\Omega}$

b. $L_C \geqq \dfrac{R_L}{6\pi f_{in}} = \dfrac{300}{6\pi \times 60}$

$\geqq 0.262\text{ H} = \mathbf{262\ mH}$

c. $C = \dfrac{0.83}{\gamma L}$ (Eq. 3.10a)

$= \dfrac{0.83}{0.02 \times 0.262}$

$= \mathbf{158\ \mu F}$

An L section filter provides a dc voltage of 30 V and a dc current of 15 mA. The choke has an inductance of 2.5 H, and C = 10 μ F. The rectifier is a full-wave circuit operating at an input frequency of 50 Hz. Is the inductance above the critical value? (R3-10)

For the circuit in (R3-10), at what frequency does the inductance cease to be critical? (R3-11)

LABORATORY EXERCISE 3-1: RC POWER SUPPLY FILTER

PURPOSE

- To observe the effects of an RC filter for a half-wave rectifier circuit and a full-wave rectifier circuit.
- To observe the ripple voltage across an RC filter.
- To use the Schade curves to verify the data observed.

MATERIALS

1 Oscilloscope
1 Voltage transformer, stepdown, with center tap, Triad No. F-54X, primary 117 V, secondary 35 V CT
1 Vacuum-tube voltmeter (VTVM)
1 Ammeter, dc, 0 to 10 mA
1 Power source, ac, 117 V, 60 Hz
2 Diodes, HEP170 suggested
1 Resistor, 10 k Ω, 1 W
1 Resistor, 2.5 kΩ, 1 W
1 Capacitor, 40 μ F, 450 V

PROCEDURE

PART I.

A. Connect the half-wave rectifier circuit as shown in figure 3-16.

B. Using the VTVM, measure the following voltages:

1. $V_o = V_{dc} = ?$
2. $V_s = V_{rms} = ?$
3. $V_{pri} = V_p = ?$

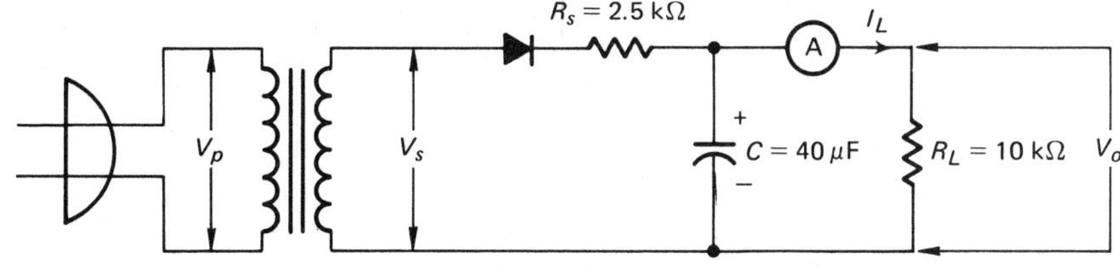

FIG. 3-16 HALF-WAVE RECTIFIER WITH AN RC FILTER

C. Using the oscilloscope, measure the following peak-to-peak voltages and draw them carefully on graph paper (use graph paper with a grid of four squares per inch).

1. The voltage across R_S.
2. The voltage across the diode.
3. The ripple voltage (the voltage across R_L).

D. Measure I_L with a dc ammeter.

E. Using the method which refers to the curves of figures 3-6, 3-7, 3-8, and 3-9, verify the data observed for the half-wave rectifier.

F. Calculate the peak and average diode currents using the Schade curves for the half-wave rectifier.

G. 1. Calculate the peak diode current using the data obtained in this exercise.

2. Compare the values of the peak current for steps F and G.1.

H. Calculate the ripple factor (γ). If the ripple voltage has a triangle (sawtooth) waveshape, use the following equation:

$$V_{r\,(rms)} = \frac{V_{r\,(p\text{-}p)}}{3.464}$$

where $V_{r\,(p\text{-}p)}$ = the peak-to-peak ripple voltage

$V_{r\,(rms)}$ = the rms ripple voltage for a sawtooth waveshape derived in reference No. 43 in the Bibliography (see the Appendix)

PART II.

A. Connect the full-wave rectifier circuit as shown in figure 3-17.

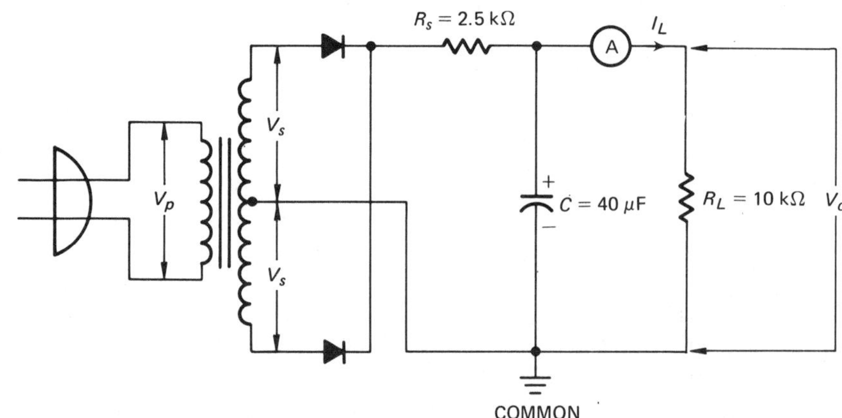

FIG. 3-17 FULL-WAVE RECTIFIER CIRCUIT WITH AN RC FILTER

B. Using the VTVM, measure the following voltages:
   1. $V_o = V_{dc} = ?$
   2. $V_s = V_{rms} = ?$
   3. $V_{pri} = V_p = ?$

C. Using the oscilloscope, measure the following peak-to-peak voltages and draw them carefully on graph paper (use paper with a grid of four squares per inch).
   1. The voltage across R_S.
   2. The voltage across the diodes.
   3. The ripple voltage (the voltage across R_L).

D. Measure I_L with a dc ammeter.

E. Repeat step E of Part I for the full-wave rectifier.

F. Repeat step F of Part I for the full-wave rectifier.

G. Repeat steps G.1. and G.2. of Part I for the full-wave rectifier.

LABORATORY EXERCISE 3-2: VOLTAGE DOUBLER

PURPOSE

- To observe the effects of a voltage doubler circuit
- To observe the effects of a choke in a power supply filter circuit

MATERIALS

1 Oscilloscope
1 Voltage transfomer, stepdown, with center tap, Triad No. F-54X, primary 117 V, secondary 35 V CT
1 Vacuum-tube voltmeter (VTVM)
4 Diodes, HEP170 suggested
3 Capacitors, 40 μF, 450 V
1 Potentiometer, 10 k Ω, 5 W
1 Choke (Inductor), 8 H (or larger)

PROCEDURE

A. Connect the voltage doubler circuit as shown in figure 3-18.

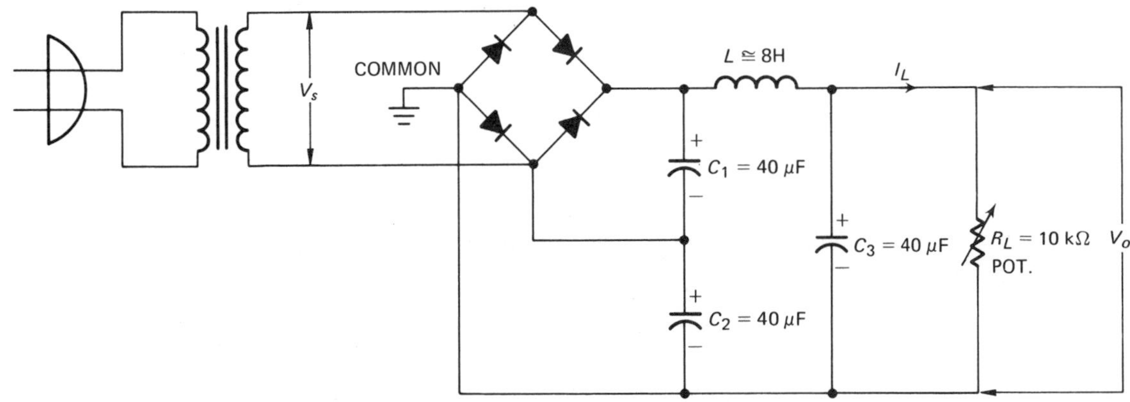

FIG. 3-18 VOLTAGE DOUBLER CIRCUIT

B. Measure $V_{S\ (rms)}$ using the VTVM.

C. Adjust R_L to the values shown below and measure V_o using the VTVM. Record the values for V_o.

R_L in kΩ	V_o in volts	R_L in kΩ	V_o in volts
0.4		1.5	
0.6		2.0	
1.0		3.0	

D. Using the oscilloscope, measure the peak-to-peak voltage across R_L. Draw this voltage carefully on graph paper (use paper with a grid of four squares per inch).

E. Calculate the dc load current, I_L, for each of the six sets of values in step C.

F. Draw a graph of V_o vs I_L.

G. Find the ripple factor γ.

H. Why is the circuit in this experiment called a voltage doubler?

EXTENDED STUDY TOPICS

1. An RC filter circuit for a half-wave rectifier requires a 5% ripple factor using a 60-Hz source. The load resistance is 10,000 ohms. Find the capacitor value.
2. An RC filter circuit for a full-wave rectifier requires a 3% ripple factor using a 400-Hz source. The dc load current is 1 mA and the dc load voltage is 7.5 V. Find the capacitance value.
3. A 2-μF capacitor is used in an RC filter circuit. The dc voltage is 10 V and the dc current is 2.5 mA. The input sine wave has a frequency of 60 Hz. Find the %γ for a half-wave rectifier circuit and a full-wave rectifier circuit.
4. A dc power supply has an open circuit voltage of 50 V. When the full-load current is drawn, the output voltage drops to 40 V. Find % V.R.
5. An L section filter is connected to a half-wave rectifier. For the filter, R_C = 200 Ω, L = 5 H, C = 10 μF, R_L = 1 kΩ, and the maximum rectifier voltage V_M = 200 V. Find the dc load voltage and the dc load current.
6. Repeat Problem 5 for a full-wave rectifier.
7. An L section filter is connected to a half-wave rectifier. The maximum rectifier voltage is V_M = 150 V, and the sine-wave input frequency is 60 Hz. The filter has the following parameters: L = 10 H, R_C = 400 Ω, C = 50 μF, and R_L = 4 kΩ. Find the dc output voltage, the ripple voltage, and the % ripple factor.

8. Repeat Problem 7 for a full-wave rectifier.

9. Find the value of capacitance for an L section filter if the inductance of the choke is 0.5 H, and the % ripple factor is 0.25%. The filter is connected to a half-wave rectifier.

10. Repeat Problem 9 for a full-wave rectifier.

11. Design an L section filter connected to a full-wave rectifier circuit to provide a dc current of 100 mA at a dc voltage of 75 V with a ripple of 1%.

Zener diodes

OBJECTIVES

After studying this unit, the student will be able to discuss and demonstrate an understanding of the basic principles of:

- The Zener diode voltage breakdown region
- Zener diode specifications
- The analysis and design of simple Zener diode voltage regulator circuits.

ZENER DIODE THEORY

When reverse bias is applied to a junction diode, a small amount of leakage current flows. If the reverse bias is large enough, a phenomenon known as *junction breakdown* or *Zener breakdown* occurs. Figure 4-1 shows Zener breakdown (large increase in current) at 10 V.

A generally accepted theory which explains why Zener voltage breakdown occurs at specified voltages is the *avalanche* theory. Avalanche breakdown occurs when holes or electrons are accelerated to very high velocities by an electric field. The electric field results when a reverse bias, usually more than five volts, is applied to the diode. The presence of the field brings about collisions between the atoms with resulting ionization of the atoms and formation of hole-electron pairs. The just formed hole-electron pairs, in turn, are accelerated and more ionizations occur, more hole-electron pairs are formed, and as a result, more current exists. The increase in current will occur suddenly at a particular voltage called the Zener voltage, or breakdown voltage.

When Zener breakdown occurs at a reverse bias of less than 5 volts, this breakdown is probably due to another phenomenon discovered by Zener. His theory states that the voltage breakdown occurs when the electric field across a narrow space charge region is strong enough to pull valence electrons from their covalent bonds. As a result, extra conduction electrons and holes are produced. These additional carriers drift across the junction under the influence of the reverse bias voltage and cause a rapid increase in the reverse current with a resulting Zener breakdown.

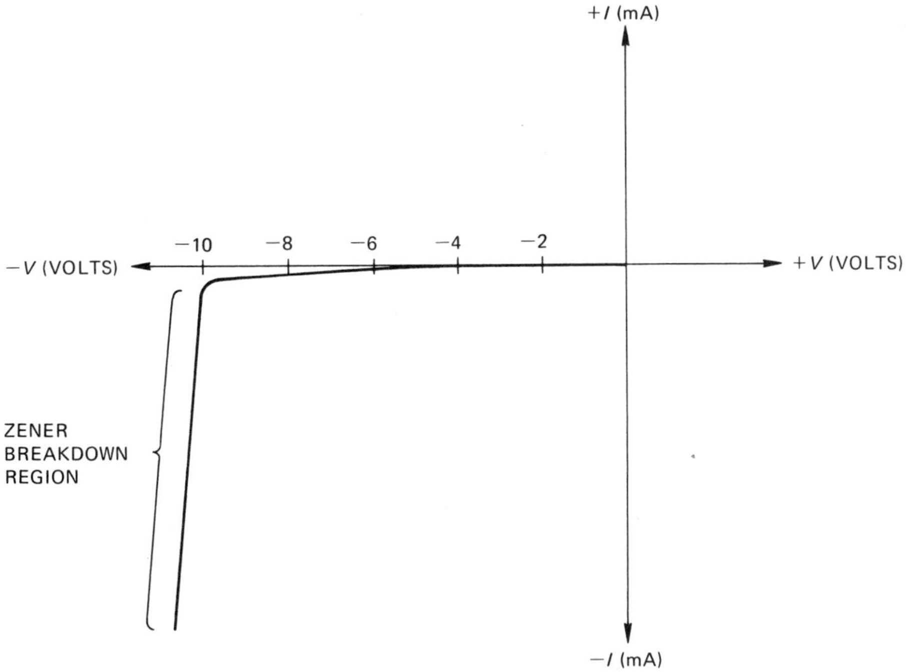

FIG. 4-1 ZENER BREAKDOWN REGION

A diode is manufactured to operate at a specific Zener breakdown voltage by varying the amount of doping and the geometry of the junction. Actually, the variation of these factors in the manufacture of the diode controls both the Zener voltage and the reverse current capabilities in the Zener voltage region.

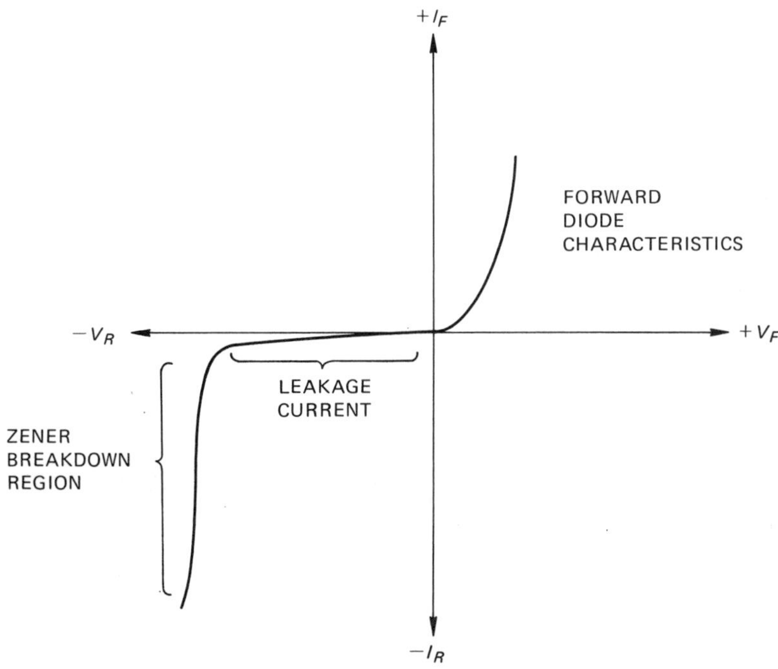

FIG. 4-2 ZENER DIODE CHARACTERISTIC CURVE

The name Zener diode is given to diodes which possess the characteristics shown in figure 4-2: the forward diode characteristics, the reverse leakage current, and the Zener breakdown voltage. Another type of diode which possesses only the Zener breakdown voltage characteristic is called the *reference* or *breakdown* diode. However, in actual use, the name Zener diode has become synonymous with the reference or breakdown diode.

Figure 4-3 shows the schematic symbols used to designate a Zener diode.

Name the two types of phenomena which cause Zener breakdown. (R4-1)

Which phenomenon causes Zener breakdown at lower voltages? Zener breakdown at higher voltages? (R4-2)

What is the difference between Zener diodes and reference diodes? (R4-3)

FIG. 4-3 TWO SYMBOLS USED TO REPRESENT ZENER DIODES

ZENER DIODE RATINGS AND SPECIFICATIONS

A typical volt-ampere Zener diode characteristic curve is shown in figure 4-4. Figure 4-5 is a reproduction of a specification sheet which gives the following important information about the specific Zener diode.

P_T = the wattage rating of the Zener diode, usually found on the front of the specification sheet

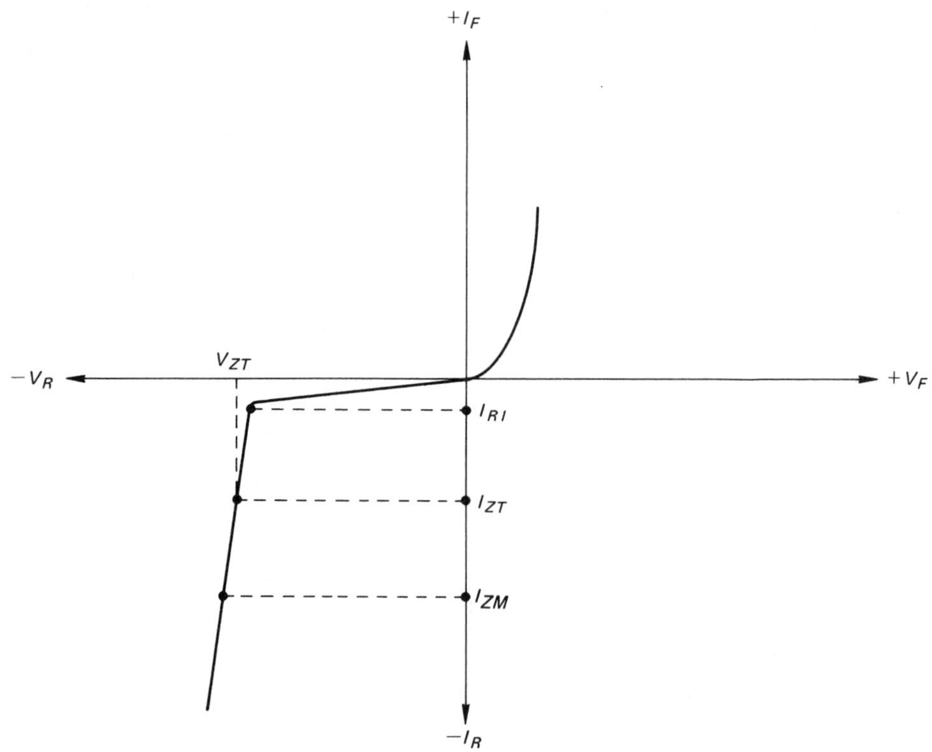

FIG. 4-4 ZENER DIODE CURVE ILLUSTRATING THE PARAMETERS V_{ZT}, I_{R1}, I_{ZT}, AND I_{ZM}

avalanche diodes

These 1 watt Avalanche (Zener) Diodes utilize the 1N2610 flangeless package. This small package is a tried and proved design with an exceptional reliability history over the past several years. The experience of many years production has been utilized in the junction design of this Zener device. The design features low junction temperatures under operating conditions. The design incorporates know-how which promotes very good voltage regulation, and voltage stability over a long period of time under the severest environmental conditions.

- **Cool Operation for Improved Characteristics and Long Life**
- **All Welded Package—Within EIA Registered D013 Outline.**
- **Operating Reliability Under Stringent Environmental Conditions**
- **Axial Leads for Ease of Mounting**

1 WATT AVALANCHE (ZENER) DIODES

SERIES
1N1765-1N1776
1N1765A-1N1776A
Z4X5.1B, Z4X5.1A
Z4X14B, Z4X14A

MAXIMUM ALLOWABLE RATINGS

Power	1 Watt
Non-recurrent Surge Rating in Zener Direction at 175°C Junction (0.001 <t<0.008 Sec.)	25 milliwatt—Sec.
Non-recurrent Surge Rating in Forward Direction at 175°C Junction (0.001 <t<0.008 Sec.)	0.50 Amp² sec.
Storage Temperature	−65°C to 175°C
Operating Junction Temperature	−65°C to 175°C

CHARACTERISTICS

Voltage	See Rating Table
Current	See Rating Table
Instantaneous Forward Drop at 0.5 Amps, 25°C (for other values—see figure 7)	Maximum = 1.1 Volts Typical = 1.0 Volts
Transient Thermal Resistance	See Figure 6
Steady State Thermal Resistance from Junction to Case	Maximum = 18.5°C/Watt Minimum = 9.7°C/Watt
Steady State Thermal Resistance from Case to Ambient (Depends on Air, Velocity, Air Temperature and temperature of surrounding surfaces.)	Typical = 60°C/Watt

FIG. 4-5

SPECIFICATIONS ON PAGES 65-68 COURTESY OF GENERAL ELECTRIC SEMICONDUCTOR PRODUCTS DEPARTMENT, SYRACUSE, N.Y.

RATING TABLE

General Electric Type Number	V_{ZT} @ I_{ZT} ZENER VOLTAGE Volts @ 25°C			Test Current I_{ZT} mA	Maximum Dynamic Impedance Z_{ZT} @ I_{ZT} @ 25°C Ohms	Maximum Leakage Current I_s @ V_s @ 25°C		Typical Temp. Coefficient %/°C	Typical Voltage Regulation ΔV_Z Volts	MAXIMUM ZENER CURRENT I_{ZM} — mA		
	10% Min.	Nominal	10% Max.			V_s Volts	I_s mA			TA = 50°C Maximum	TA = 100°C Maximum	TA = 150°C Maximum
Z4X5.1B	4.6	5.1	5.6	100	7.0	4.1	15.00	0.013	0.11	160	96	32
1N1765	5.0	5.6	6.2	100	1.2	4.5	5.00	0.021	0.14	150	88	29
1N1766	5.6	6.2	6.8	100	1.5	5.0	1.70	0.030	0.16	130	79	26
1N1767	6.1	6.8	7.5	100	1.7	5.4	0.96	0.037	0.20	120	72	24
1N1768	6.8	7.5	8.3	100	2.1	6.0	0.75	0.044	0.24	109	65	22
1N1769	7.4	8.2	9.0	100	2.4	6.6	0.64	0.050	0.28	100	60	20
1N1770	8.2	9.1	10.0	50	3.0	7.3	0.52	0.056	0.34	90	54	18
1N1771	9.0	10.0	11.0	50	3.5	8.0	0.44	0.062	0.41	82	49	16
1N1772	9.9	11.0	12.1	50	4.2	8.8	0.38	0.067	0.48	74	45	15
1N1773	10.8	12.0	13.2	50	5.0	9.6	0.34	0.071	0.57	68	41	14
1N1774	11.7	13.0	14.3	50	5.8	10.4	0.31	0.074	0.66	63	38	13
Z4X14B	12.6	14.0	15.4	50	6.6	11.2	0.29	0.077	0.75	58	35	12
1N1775	13.5	15.0	16.5	50	7.6	12.0	0.26	0.080	0.86	54	33	11
1N1776	14.4	16.0	17.6	50	8.6	12.8	0.24	0.082	0.97	51	31	10

NOTE: Standard types are supplied to ±10% of voltage values listed.
For 5% tolerance:
1. Add "A" suffix to 1N number, i.e. 1N1776A
2. Change "B" suffix on Z4X number to "A," i.e. Z4X5.1A

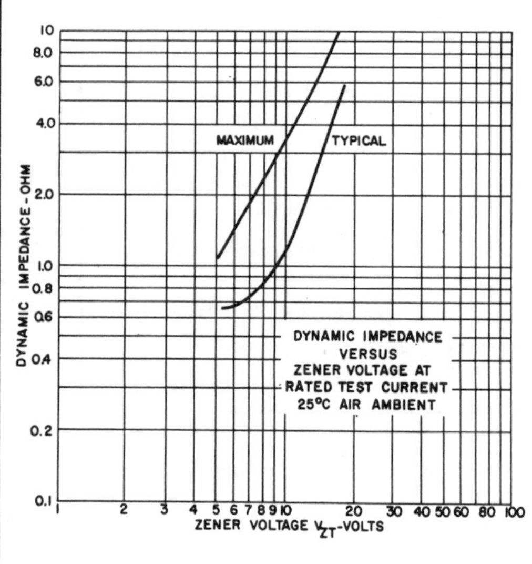

1. DYNAMIC IMPEDANCE

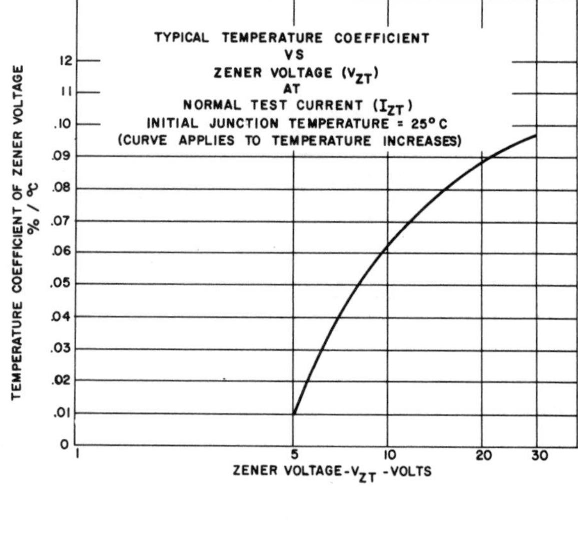

2. TYPICAL TEMPERATURE COEFFICIENT

FIG. 4-5 (Continued)

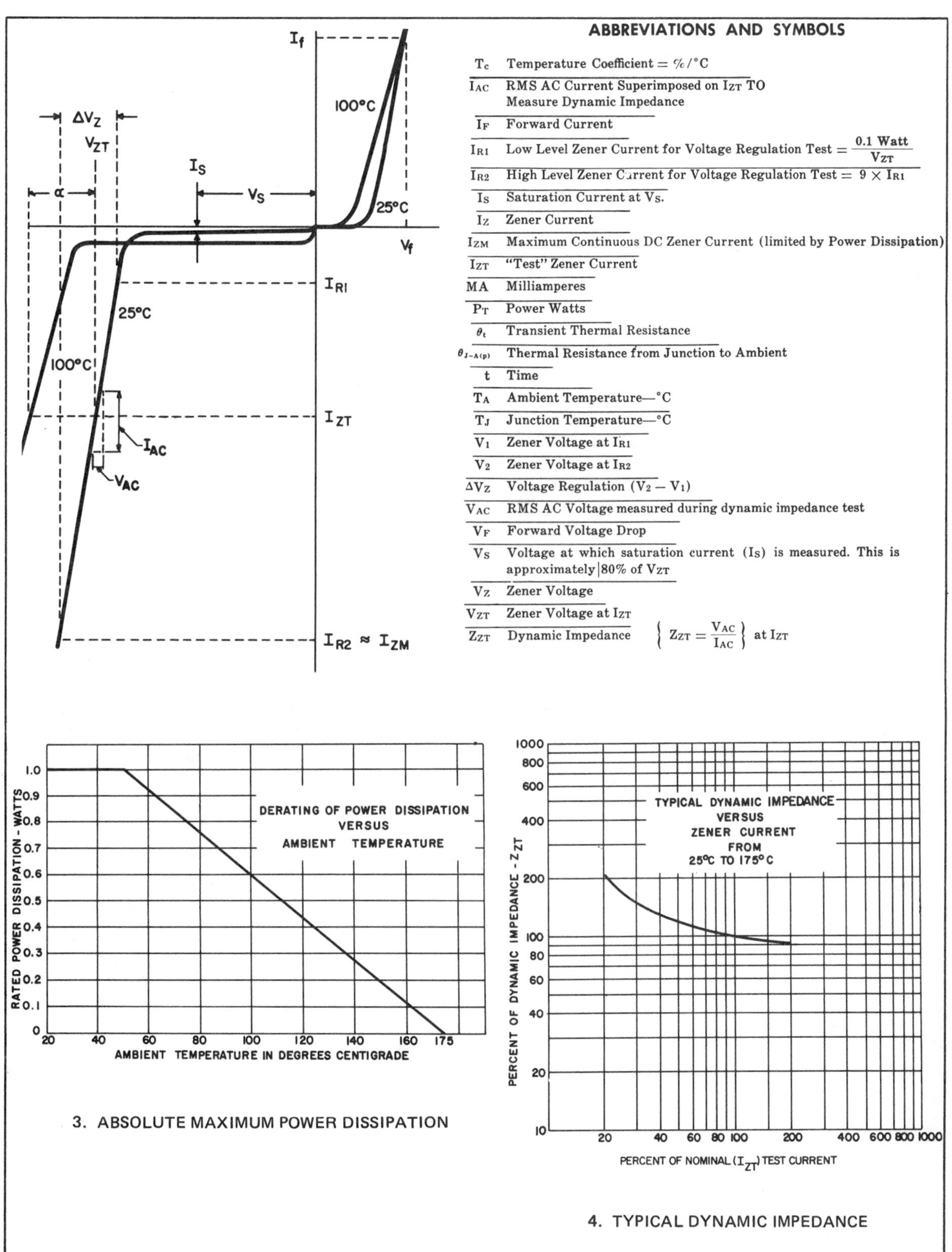

ABBREVIATIONS AND SYMBOLS

T_c	Temperature Coefficient = %/°C
I_{AC}	RMS AC Current Superimposed on I_{ZT} TO Measure Dynamic Impedance
I_F	Forward Current
I_{R1}	Low Level Zener Current for Voltage Regulation Test = $\frac{0.1 \text{ Watt}}{V_{ZT}}$
I_{R2}	High Level Zener Current for Voltage Regulation Test = $9 \times I_{R1}$
I_S	Saturation Current at V_S.
I_Z	Zener Current
I_{ZM}	Maximum Continuous DC Zener Current (limited by Power Dissipation)
I_{ZT}	"Test" Zener Current
MA	Milliamperes
P_T	Power Watts
θ_t	Transient Thermal Resistance
$\theta_{J-A(p)}$	Thermal Resistance from Junction to Ambient
t	Time
T_A	Ambient Temperature—°C
T_J	Junction Temperature—°C
V_1	Zener Voltage at I_{R1}
V_2	Zener Voltage at I_{R2}
ΔV_Z	Voltage Regulation $(V_2 - V_1)$
V_{AC}	RMS AC Voltage measured during dynamic impedance test
V_F	Forward Voltage Drop
V_S	Voltage at which saturation current (I_S) is measured. This is approximately 80% of V_{ZT}
V_Z	Zener Voltage
V_{ZT}	Zener Voltage at I_{ZT}
Z_{ZT}	Dynamic Impedance $\left\{ Z_{ZT} = \frac{V_{AC}}{I_{AC}} \right\}$ at I_{ZT}

3. ABSOLUTE MAXIMUM POWER DISSIPATION

4. TYPICAL DYNAMIC IMPEDANCE

FIG. 4-5 (Continued)

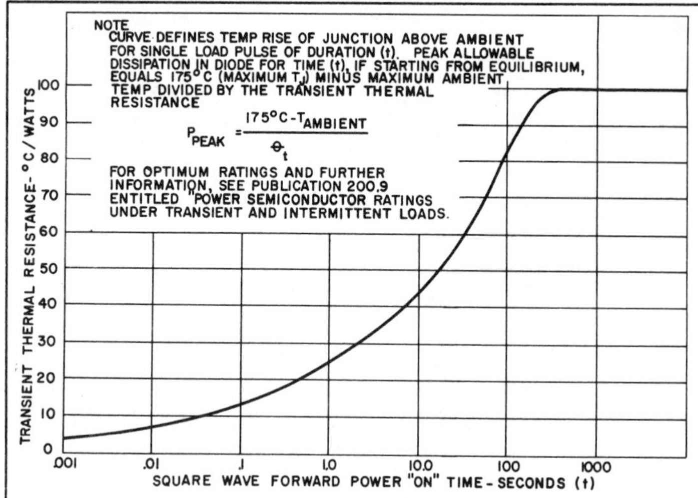

5. MAXIMUM JUNCTION TO AMBIENT TRANSIENT THERMAL RESISTANCE (No Heat Sink)

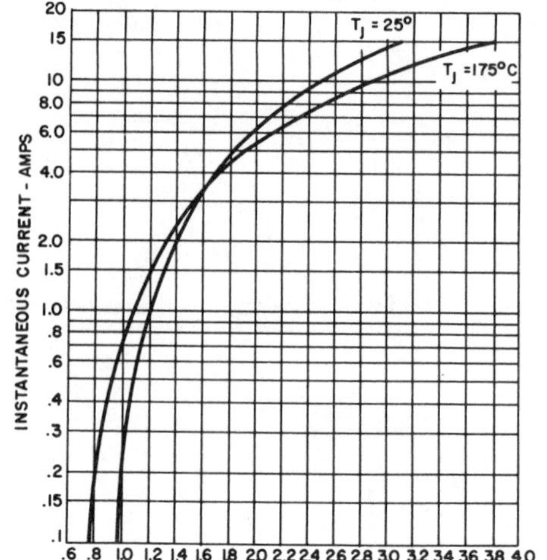

6. MAXIMUM INSTANTANEOUS FORWARD VOLTAGE

STANDARD TEST CONDITIONS

VOLTAGE CLASSIFICATION applies under conditions of Thermal equilibrium in a free convection Air Ambient held at 25° ± 1°C, with the Diode mounted with 3/8 inch of its own flexible lead between the mounting and the Diode body.

DYNAMIC IMPEDANCE is determined by superimposing 10% of test current listed (60 cycle, AC) on DC test current. For typical values of Dynamic Impedance and temperature coefficients, see attached curves.

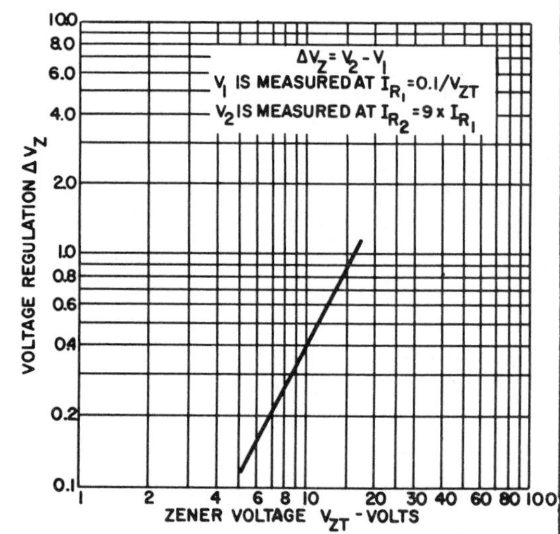

7. TYPICAL VOLTAGE REGULATION

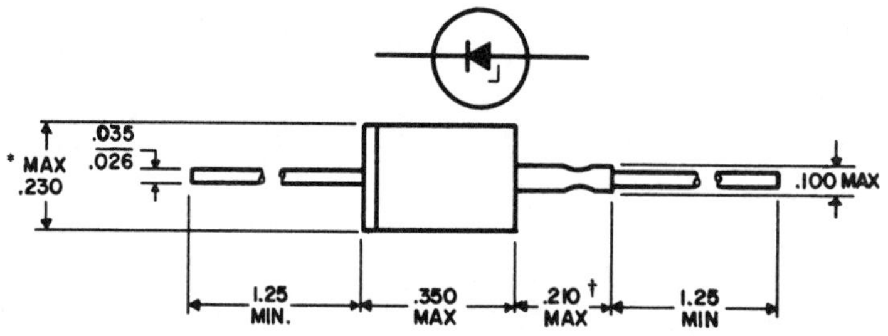

Complies with JEDEC DO-13 outline.
*Registration Data on 1N1765 series indicates a small flange with max. .240 and a barrel diameter of max. .200. This flangeless housing .230 diameter fits within the registered flange max. diameter.
†Registration Data gives this dimension as .135.

FIG. 4-5 (Continued)

V_{ZT} = the nominal Zener breakdown voltage at a specific current I_{ZT}

I_{ZT} = the test Zener current near the middle of the operating current range

I_{ZM} = the maximum continuous dc Zener current (limited by the power dissipation P_T)

I_{R1} = the minimum Zener current

I_{R2} = a high level Zener current used for voltage regulation tests

Z_{ZT} = the dynamic impedance of a Zener diode

Three of the preceding parameters are expressed by the following equations.

$$I_{ZM} = \frac{P_T}{V_{ZT}} \qquad \text{Eq. 4.1}$$

$$I_{R1} = \frac{0.1 \times P_T}{V_{ZT}} \qquad \text{Eq. 4.2}$$

$$I_{R2} = 9 \times I_{R1} \qquad \text{Eq. 4.3}$$

Equation 4.1 indicates that the maximum Zener current (I_{ZM}) is obtained by dividing the power rating of the Zener diode by the voltage rating of the Zener. If the current through the Zener exceeds this value, the diode will be destroyed.

Equation 4.2 expresses the minumum operating Zener current (I_{R1}) in terms of the power rating and the voltage rating of the Zener diode. If the Zener current falls below this value, the diode will not regulate. Equation 4.2 is a close estimate only for a particular Zener diode.

Equation 4.3 for the maximum *operating* current (I_{R2}) gives the approximate value supplied by the Zener diode manufacturer. This is the maximum current at which a Zener diode should be operated and is equal to nine times the minimum Zener current (I_{R1}).

Zener diodes are temperature dependent. Those diodes which break down due to the avalanche phenomenon have a positive temperature coefficient K_T. Zener diodes which break down due to the junction phenomenon have a negative temperature coefficient K_T. The temperature coefficient K_T, which is stated on the specification sheets published by diode manufacturers, indicates the percentage change in the nominal Zener voltage, V_{ZT}. Refer to the RATING TABLE in figure 4-5 and locate the column titled "Typical Temp. Coefficient %/°C". The Zener diodes covered by the specification sheet shown in figure 4-5 are avalanche diodes; therefore, K_T is positive. Problem 1 shows how the nominal Zener voltage V_{ZT} is adjusted for temperature changes.

PROBLEM 1

A Zener diode, type number 1N1769, is used in a situation where the temperature may increase from 25°C to 125°C. Find the value of V_{ZT} at 100°C (V'_{ZT}).

Locate the values for V_{ZT} and K_T from figure 4-5 for the type 1N1769 diode.

$$V_{ZT} = 8.2\text{ V at } 25^\circ\text{C}$$

$$K_T = \frac{0.05\%}{^\circ\text{C}}$$

K_T gives the percentage change per degree C; therefore, use K_T first to calculate the voltage change per degree C.

$$\frac{\Delta V_{ZT}}{^\circ\text{C}} = \frac{0.0005}{^\circ\text{C}} \times 8.2\text{ V}$$

$$= 4.1\ \frac{\text{mV}}{^\circ\text{C}}$$

Next find the temperature change.

$$\Delta T = (125 - 25)\ ^\circ\text{C}$$

$$= 100^\circ\text{C}$$

Multiply the temperature change by the voltage change per °C to find the voltage change.

$$\Delta V_{ZT} = \frac{\Delta V_{ZT}}{°C} \times T$$

$$= 4.1 \frac{mV}{°C} \times 100°C$$

$$= 410 \text{ mV}$$

$$= 0.41 \text{ V}$$

Now add the voltage change to the nominal V_{ZT}.

$$V'_{ZT} = V_{ZT} + \Delta V_{ZT}$$

$$= 8.2V + 0.41 \text{ V}$$

$$= \mathbf{8.61 \text{ V}}$$

A Zener diode has the following ratings: P_T = 10W, and V_{ZT} = 20 V. Find I_{ZT}, I_{R1}, and I_{R2}. (R4-4)

For (R4-4), find the Zener impedance at I_{R1} and I_{R2} if the Zener breakdown voltage varies ± 10% from V_{ZT} (R4-5)

For a type 1N1775 Zener diode, find the value of V_{ZT} at 75°C. (R4-6)

Repeat (R4-6) for a type 1N1765 Zener diode. (R4-7)

ZENER DIODE VOLTAGE REGULATOR CIRCUITS

The most popular use of the Zener diode is in a voltage regulator circuit, as shown in figure 4-6. A voltage regulator circuit will maintain a steady dc voltage across some load (R_L) over a range of operating currents. The voltage source V_{oc} is *un-*

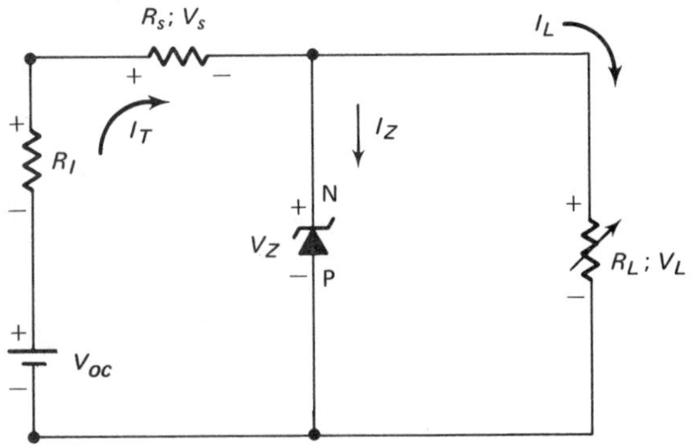

FIG. 4-6 TYPICAL ZENER DIODE VOLTAGE REGULATOR CIRCUIT

regulated; that is, it fluctuates depending upon its internal R_1 and the total current flow. To maintain a constant voltage V_{ZT} across the load R_L, the Zener diode is used. The diode is operated in its breakdown region where its voltage (V_{ZT}) varies only slightly over a wide range of currents (I_{R1} to I_{ZM}). The resistor R_S in the circuit keeps the Zener diode within its voltage and current limits.

The operation of the voltage regulator circuit in figure 4-6 is best described by assuming the following values for the circuit parameters. Let V_{oc} = 15 V, R_I = 100Ω, R_S = 100 Ω, and V_L = 5 V. First of all, assume that the load is an open circuit; that is, $R_L = \infty$ and I_L = 0. Applying Kirchhoff's current law to the circuit in the figure, an expression for the total current is obtained.

$$\mathbf{I_T = I_L + I_Z} \qquad \textbf{Eq. 4.4}$$

Since I_L = 0, then $I_T = I_Z$. I_T is determined by the use of Kirchhoff's voltage law.

$$\mathbf{V_{oc} = I_T (R_I + R_S) + V_Z} \qquad \textbf{Eq. 4.6}$$

$$\mathbf{V_Z = V_L}$$

By substituting the known values into Eq. 4.5, I_T can be determined.

$$15 \text{ V} = I_T (100\Omega + 100\Omega) + 5 \text{ V}$$

$$I_T = \frac{15 \text{ V} - 5 \text{ V}}{200 \Omega}$$

$$= \frac{10\ V}{200\ V}$$

$$= 0.05A = 50\ mA$$

Thus, when the load is open, the current through the Zener diode is 50 mA. This current is the maximum current flow through the Zener diode. The I_{ZM} rating of the Zener diode should be greater than 50 mA to insure safe operation. In other words, the maximum Zener diode current flow should be calculated using the minimum load current.

If a finite value of load resistance is now connected to the circuit, this load will draw a current whose magnitude depends on the load value. This current is subtracted from the 50 mA being supplied by the voltage source; thus, less current is available for the Zener diode.

If a load current of 10 mA is assumed, then the load resistance $R_L = V_L/I_L = 5V/10\ mA = 500\ \Omega$. Since the total current remains *constant* at 50 mA, the current available for the Zener diode is $I_Z = I_T - I_L = 50mA - 10\ mA = 40\ mA$. The I_{R1} rating of the Zener diode should be less than 40 mA to insure that the Zener diode works properly. In other words, if the Zener diode is to operate in its voltage breakdown region, the minimum diode current flow must be calculated when the load current is maximum. In addition, I_{R1} must not exceed this minimum Zener diode current flow.

Problem 2 and 3 illustrate the processes of analyzing and designing Zener diode voltage regulator circuits.

PROBLEM 2.

A 15-V unregulated dc power supply has an internal resistance of 0 Ω.

Design a voltage regulator circuit for a load voltage of 6.2 V, and current variations of 5 mA to 50 mA.

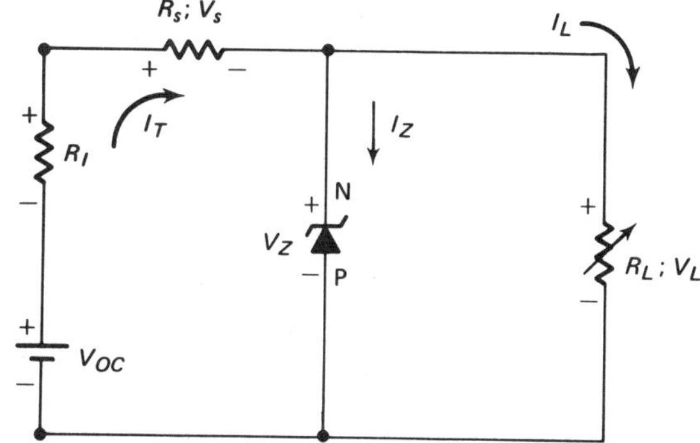

Before starting the analysis, draw a fully labeled schematic of a voltage regulator circuit.

a. From the specification given in figure 4-5, choose a diode that will maintain a load voltage of 6.2 V. Copy the relevant data for this diode. The diode selected to satisfy the 6.2 V requirement should be the type 1N1766 diode for which $V_{ZT} = 6.2\ V$, $I_{ZT} = 100\ mA$, and $P_T = 1\ W$.

b. Calculate I_{ZM} and I_{R1}:

$$I_{ZM} = \frac{P_T}{V_{ZT}}$$

$$= \frac{1\ W}{6.2\ V}$$

$$= 0.161\ A = \mathbf{161\ mA}$$

$$I_{R1} = \frac{0.1\ P_T}{V_{ZT}}$$

$$= \frac{0.1 \times 1W}{6.2\ V}$$

$$= \mathbf{16.1\ mA}$$

c. Using Kirchhoff's current law, find I_T.

$$I_T = I_L + I_Z$$

d. When the maximum current flows in the load, then the Zener diode minimum current $I_Z = I_{R1}$

$$I_T = I_{L\,(max)} + I_{R1}$$
$$= 50\text{ mA} + 16.1\text{ mA}$$
$$= \mathbf{66.1\text{ mA}}$$

e. When the minimum current flows in the load, the value of the current I_{ZA} for the Zener diode must be such that the total current equals 66.1 mA.

$$I_T = I_{L\,(min)} + I_{ZA}$$
$$66.1\text{ mA} = 5\text{ mA} + I_{ZA}$$
$$I_{ZA} = \mathbf{61.1\text{ mA}}$$

I_{ZA} in this case is much less than I_{ZM}; therefore, the 1N1766 diode can be used in this circuit.

f. Calculate R_S

$$R_S = \frac{V_S}{I_T}$$

$$V_S = V_{oc} - V_L$$
$$= 15\text{ V} - 6.2\text{ V}$$
$$= 8.8\text{ V}$$

$$R_S = \frac{8.8\text{ V}}{66.1\text{ mA}}$$
$$= 0.133 = \mathbf{133\ \Omega}$$

g. Calculate the power rating of R_S.

$$P_S = V_S I_T$$
$$= 8.8\text{ V} \times 66.1\text{ mA}$$
$$= 581.7\text{ mW} = 0.5817\text{ W}$$

A 1-W resistor is adequate for R_S.

PROBLEM 3.

A 50-V unregulated dc power supply has a value of $R_1 = 100\ \Omega$. The load voltage is 10 V and the load current is 40 mA. Design a voltage regulator circuit and specify the Zener diode to be used, R_S, and P_S.

a. Refer to figure 4-5 and select a Zener diode for which $V_{ZT} = V_L = 10$ V. Such a diode is type 1N1771. A value for P_T is also known: $P_T = 1$ W

b. Calculate the values of I_{ZM} and I_{R1}.

$$I_{ZM} = \frac{P_T}{V_{ZT}}$$
$$= \frac{1\text{ W}}{10\text{ V}}$$
$$= \mathbf{100\text{ mA}}$$

$$I_{R1} = \frac{0.1\ P_T}{V_{ZT}} = 0.1\ I_{ZM}$$
$$= 0.1 \times 100\text{ mA} = \mathbf{10\text{ mA}}$$

c. To achieve the best operation, the Zener diode must operate in the Zener region between the current values of I_{R1} and I_{ZM} when the load is at 40 mA. This intermediate value of I is called $I_{Z\,(mid)}$.

$$I_{Z\,(mid)} = I_{R1} + \frac{I_{ZM} - I_{R1}}{2}$$
$$= 10\text{ mA} + \frac{100\text{ mA} - 10\text{ mA}}{2}$$
$$= 10\text{ mA} + 45\text{ mA}$$
$$= \mathbf{55\text{ mA}}$$

d. Calculate the total current I_T.

$$I_T = I_L + I_{Z}(mid)$$
$$= 40\text{ mA} + 55\text{ mA}$$
$$= \mathbf{95\text{ mA}}$$

e. Calculate the voltage across R_S which is $V_S = I_T R_S$

$$V_{oc} = I_T R_I + I_T R_S + V_Z$$
$$50\text{ V} = (95\text{ mA} \times 100\Omega) + V_S + 10\text{ V}$$
$$V_S = 50\text{ V} - 9.5\text{ V} - 10\text{ V}$$
$$= 30.5\text{ V}$$

f. Find R_S and P_S

$$R_S = \frac{V_S}{I_T}$$

$$= \frac{30.5\ \text{V}}{95.0\ \text{mA}}$$

$$= 0.321\ \text{k}\Omega$$

$$P_S = V_S I_T$$

$$= 30.5\ \text{V} \times 95\ \text{mA}$$

$$= 2.9\ \text{W}$$

A 5-W resistor is sufficient for this circuit.

Design a voltage regulator for a 30-V unregulated dc power supply with R_I = 50 Ω. The load voltage must be 6.2 V and the load current is 50 mA. Find the proper Zener diode to be used, R_S, and P_S. (R4-8)

LABORATORY EXERCISE 4-1: ZENER DIODE VOLTAGE REGULATOR

PURPOSE

- To determine the characteristics of a Zener diode
- To observe how the Zener diode is used as a voltage regulator

MATERIALS

1 Power supply, dc, variable
1 Vacuum-tube voltmeter (VTVM)
1 Volt-ohm milliammeter (VOM)
1 Milliammeter
1 Zener diode, 1N3020, 1 W
1 Resistor, 3.3 kΩ, 1 W
1 Resistor, 500 Ω, 5 W

PROCEDURE

A. 1. Determine the *p* and *n* sides of the Zener diode.

2. Connect the circuit as shown in figure 4.7.

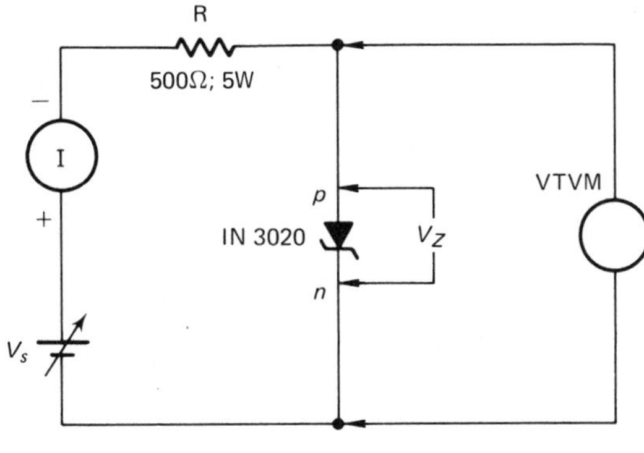

FIG. 4-7

B. 1. Adjust the Zener diode voltage (V_{ZF}) to the values shown in Table 4-1.

2. Measure and record I.

3. Calculate the resistance R_{ZF} of the Zener diode, $R_{ZF} = V_{ZF}/I$. Record the values for R_{ZF} in Table 4-1.

4. Is the Zener diode forward or reverse biased? Why?

TABLE 4-1

V_{ZF} in volts	0	0.1	0.2	0.3	0.4	0.5	0.52	0.54	0.56	0.58	0.6	0.62
I in mA												
R_{ZF} in ohms												

C. Set the power supply to zero volts.

D. Reverse the diode in the circuit of figure 4-7.

E. 1. Measure and record in Table 4-2 the current I for the voltages shown in Table 4-2.

2. Calculate the resistance R_{ZR} of the Zener diode, $R_{ZR} = V_{ZR}/I$. Record the values for R_{ZR} in Table 4-2.

3. Is the Zener diode forward or reverse biased? Why?

TABLE 4-2

V_{ZR} in Volts	I in mA	R_{ZR} in ohms	V_{ZR} in Volts	I in mA	R_{ZR} in ohms	V_{ZR} in Volts	I in mA	R_{ZR} in ohms
0.0			9.5			10.5		
1.0			9.6			10.6		
6.0			9.7			10.7		
7.0			9.8			10.8		
8.0			9.9			10.9		
8.5			10.0			11.0		
9.0			10.1			11.1		
9.2			10.2			11.2		
9.3			10.3			11.3		
9.4			10.4			11.4		

F. Draw a graph of the Zener diode in the forward and reverse directions.

G. Connect the circuit of figure 4-8 with $V_s = 0$.

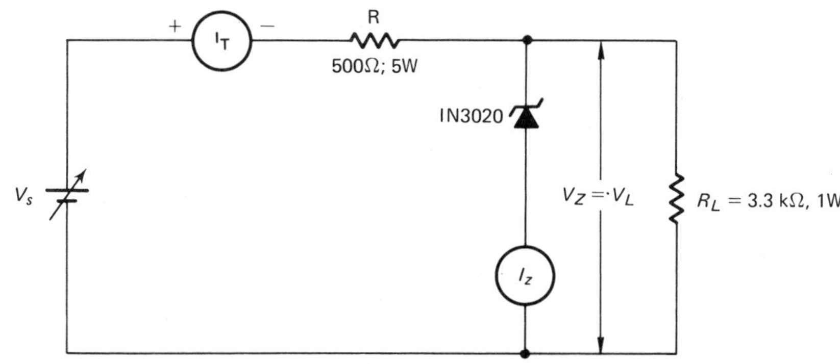

FIG. 4–8

H. 1. Slowly increase V_S until the current I_Z equals 10 mA.

2. Measure V_S, V_Z, and I_T. Record these values in Table 4-3.

I. Increase V_Z one percent (1%) above V_Z in step H.2. and repeat steps H.1. and H.2.

TABLE 4-3

	V_Z in Volts	I_Z in mA	I_T in mA	V_s in Volts
V_Z				
1 % above V_Z				
1 % below V_Z				

J. Decrease V_Z one percent (1%) below V_Z in step H.2. Repeat steps H.1. and H.2.

K. Explain how the voltage regulator circuit works from the results in steps H, I, and J.

EXTENDED STUDY TOPICS

1. For a Zener diode, type ZAX14B, find the value of V_{ZT} at 65° C.

2. A Zener diode has the following ratings:
   P_T = 1W, and V_{ZT} = 30 V.
   Find I_{ZT}, I_{R1}, and I_{R2}.

3. A voltage source has an internal resistance of 2 ohms. The open circuit voltage of the dc source is 12 V. A load is placed across the source and the terminal voltage becomes 10 V.
   Find the current and the load resistance.

4. Design a voltage regulator circuit for a load voltage of 10.8 V and load current variations of 1 mA to 15 mA, if V_{oc} = 20 V and R_I = 0.

5. For the circuit in figure 4-6, $V_{OC} = 35$ V, $R_I = 5\ \Omega$, and $R_S = 35\ \Omega$. The Zener diode has ratings of $P_T = 2$ W and $V_{ZT} = 20$ V. Find the range over which R_L can be varied.

6. For the circuit in figure 4-6, $R_I = 0\ \Omega$, $R_S = 4\ \text{k}\ \Omega$, $R_L = 10\ \text{k}\ \Omega$, and $V_{ZT} = 25$ V. The voltage V_{OC} varies between 50 V and 75 V. Find the minimum currents through the Zener diode.

7. For the circuit in figure 4-6, $R_I = 0\ \Omega$, $R_S = 3\ \Omega$, $R_L = 18\ \Omega$, $V_{ZT} = 10$ V, and $P_T = 20$ W. Find the allowable range of V_{OC}.

Junction transistor familiarization

OBJECTIVES

After studying this unit, the student will be able to discuss and demonstrate an understanding of the basic principles of:

- The operation of a PNP or NPN transistor.
- Common-base transistor configurations and common-base transistor characteristics.
- Calculating α (alpha) from the common-base transistor characteristics or from data supplied from a common-base transistor circuit.
- Connecting a common-base transistor circuit and obtaining the data that explains the characteristics of this type of circuit.

JUNCTION TRANSISTOR ACTION

Figure 5-1 shows the basic physical structure of PNP and NPN transistors. Figure 5-2 illustrates the schematic symbol used to represent each type of resistor. Note that both types of transistors have two junctions. One junction is called the *emitter-base junction* and the other is called the *collector-base junction*. Most junction transistors are identified by a type number beginning with 2N (meaning two junctions) followed by several digits. For example, a common transistor is type number 2N4425. In the symbols shown in figure 5-2, note that the arrow on the emitter of the PNP transistor *P*oints i*N*to the transistor. In addition, note that the arrow on the emitter of the NPN transistor does *N*ot *P*oint i*N*to the transistor. (These arrows point in the direction of conventional current flow. Recall that conventional current flows in the opposite direction to electron current.)

Figures 5-1 and 5-2 identify the various parts of a transistor. The *emitter* section of a transistor *emits* charge carriers. Thus, if the emitter is a *p*-type semiconductor, it emits holes. If the emitter is an *n*-type semiconductor, it emits electrons. The *base* of a transistor controls the flow of charge carriers. Physically, the base is small so that charge carriers can move quickly through it. In addition, the base is lightly doped so that a small number of charge carriers from the emitter recombine with the opposite charge carriers in the base. The action of the base will be explained later in this unit. The *collector* of a transistor *collects* charge carriers. The collector of an NPN transistor collects electrons; the collector of a PNP transistor collects holes.

Figure 5-3 will help to explain current flow through an NPN transistor. In figure 5-3b, the majority carrier electrons are to be moved from the emitter to the base. To do this, the emitter-base junction is forward biased.

The collector-base junction is reverse biased in figure 5-3c. The reverse bias permits a small *reverse leakage current* consisting of thermally generated *minority* carriers. This *collector* leakage current, I_{CBO}, flows when $I_E = 0$ and the collector-base junction is reverse biased.

Figure 5-3d illustrates the combined biasing required for the emitter-base and collector-base junctions. The phrase *transistor action* is applied to the behavior which results from this biasing. Because the emitter-base junction is forward biased, emitter electrons will move from the emitter region to the base region. Only a few of these emitter electrons will recombine with the few holes present in the lightly doped base region.

The remaining emitter electrons (approximately 95%) will move across the collector-base junction. To permit the free movement of the electrons toward the collector-base junction, the base is physically narrow and is lightly doped. The electrons in the base are pulled across the collector-base junction by the external battery connected to the junction. That is, the collector is made positive with respect to the base. Thus, the electrons are attracted to the collector, cross the collector-base junction, move through the

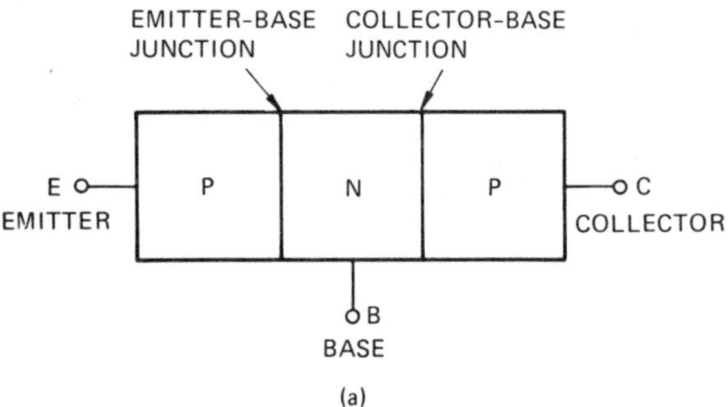

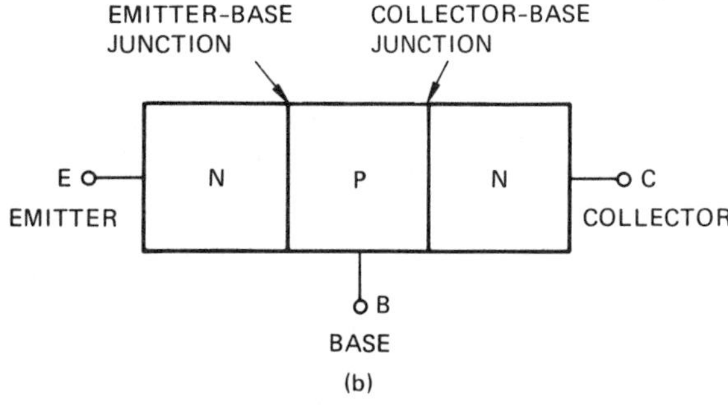

FIG. 5-1 PNP AND NPN TRANSISTORS

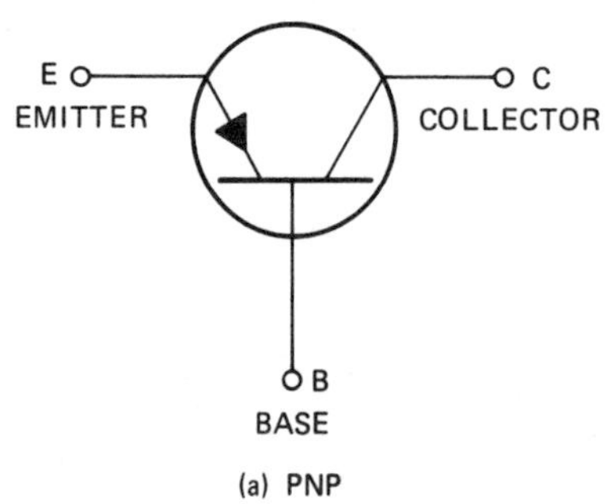

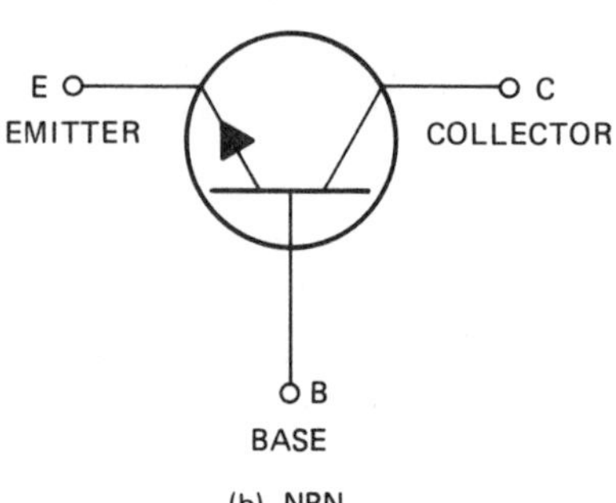

FIG. 5-2 PNP AND NPN TRANSISTOR SYMBOLS

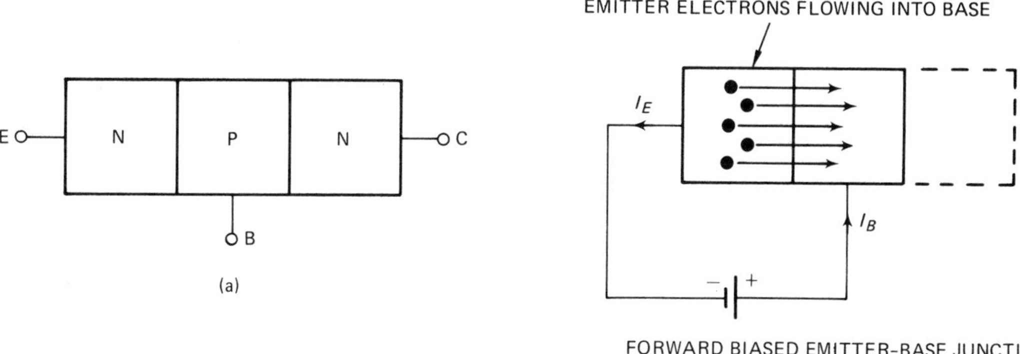

(a)

(b)

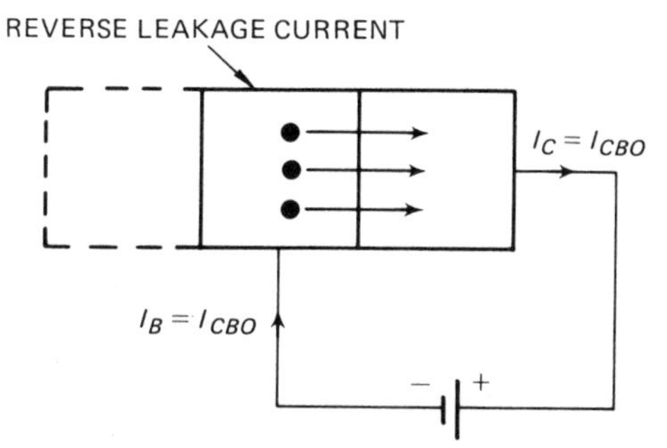

(c)

DIFFUSED EMITTER ELECTRONS (9 OUT OF 10)
IN COLLECTOR REGION

I_E

I_C

EMITTER ELECTRONS
(1 OUT OF 10) THAT
RECOMBINE IN BASE

I_B

− +

− +

FORWARD BIAS

REVERSE BIAS

(d)

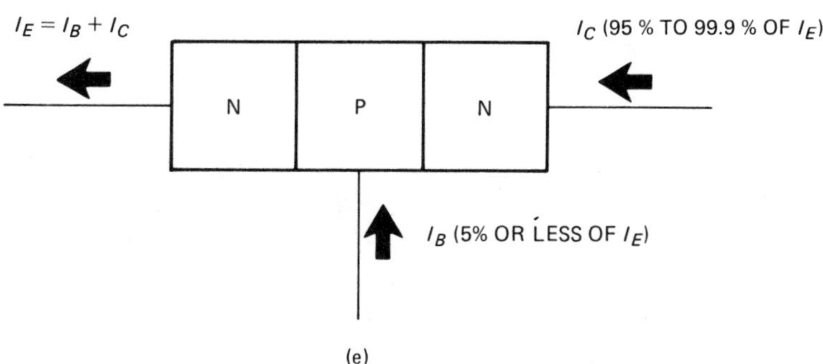

(e)

FIG. 5-3 CURRENT FLOW IN AN NPN TRANSISTOR

collector region toward the positive battery potential, and finally move through the external circuit. The approximate current values for an NPN transistor are shown in figure 5-3e.

If the potential at the collector-base junction is zero, then only some of the electrons will have enough momentum to cross the junction, move through the collector region, and move through the external circuit. If the collector is made negative with respect to the base (forward biased), then no electrons will move across the collector-base junction due to the repelling force of the negative battery potential.

The analysis for a PNP transistor is similiar to that for an NPN transistor. For the PNP transistor, however, the majority current carriers in the emitter are holes, the emitter-base junction is forward biased, and the collector-base junction is reverse biased as shown in figure 5-4a. Figure 5-4b illustrates the approximate current values for a PNP transistor.

A transistor current equation that defines the transistor action can be written with the use of Kirchhoff's current law.

$$I_E = I_B + I_C \qquad \text{Eq. 5.1}$$

where I_B = base current

I_C = collector current

I_E = emitter current

As shown in previous paragraphs, the collector current consists of two parts: the reverse leakage current I_{CBO} and the majority percentage of the emitter current which reaches the collector. Mathematically, the collector current is expressed as:

$$I_C = \alpha I_E + I_{CBO} \qquad \text{Eq. 5.2}$$

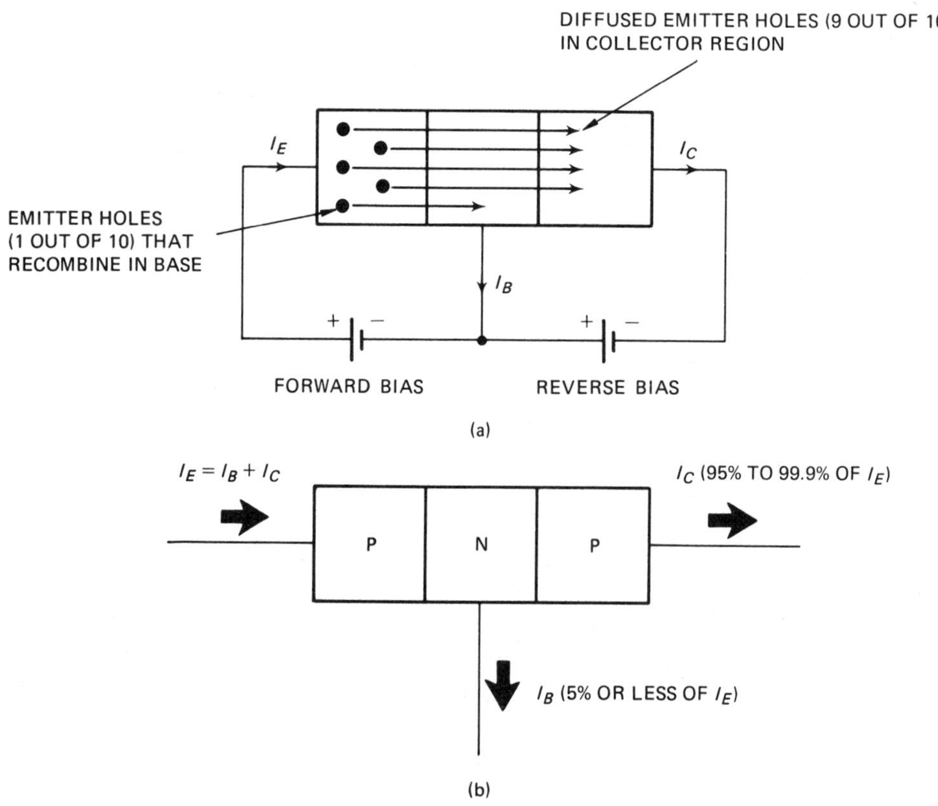

FIG. 5-4 CURRENT FLOW IN A PNP TRANSISTOR

where

I_{CBO} = reverse leakage current

I_E = emitter current

α (alpha) = the majority percentage of the emitter current which reaches the collector. This quantity is called the *common-base (CB) current gain.*

I_C = collector current

Solving for α in Eq. 5.2 yields:

$$\alpha = \frac{I_C - I_{CBO}}{I_E} \qquad \textbf{Eq. 5.3}$$

When the reverse leakage current is small compared to the collector current, Eq. 5.3 becomes:

$$\alpha = \frac{I_C}{I_E} \qquad \textbf{Eq. 5.4}$$

Values of α normally range from 0.9 to 0.999, indicating that most of the emitter current becomes collector current. The quantity α may also be referred to as α_{dc} or h_{FB}. When ac quantities are present, the CB current gain is defined as:

$$\alpha_{ac} = \left.\frac{\Delta I_C}{\Delta I_E}\right|_{V_{CB} = \text{constant value}} \qquad \textbf{Eq. 5.5}$$

where

ΔI_C = small change in collector current

ΔI_E = small change in emitter current

V_{CB} = constant collector to base voltage

α_{ac} = common base ac current gain, sometimes called h_{FB}.

I_B can be solved in terms of I_E by substituting Eq. 5.2 into 5.1 as follows:

$$I_E - I_B = \alpha I_E + I_{CBO}$$

$$I_B = (1 - \alpha) I_E - I_{CBO} \qquad \textbf{Eq. 5.6}$$

The remaining sections of this unit will cover three of the common modes of operation for a transistor, figure 5-5. For each of these modes, one of the three transistor leads will be common to both the input and the output. The common terminal designates the name of the transistor configuration. When the base is common, the result is a common base (CB) circuit; a common emitter (CE) circuit results when the emitter is common; and a common collector (CC) circuit results when the collector is common.

Draw the transistor symbol for a PNP and an NPN transistor. (R5-1)

Define the following terms: emitter, base, and collector. (R5-2)

Describe transistor action for a PNP transistor. (R5-3)

Write the transistor current equation. (R5-4)

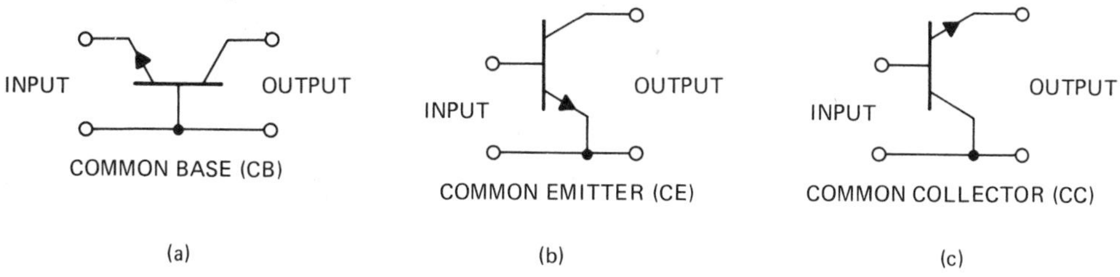

FIG. 5-5 THE THREE TRANSISTOR CONFIGURATIONS FOR AN NPN TRANSISTOR

COMMON-BASE CHARACTERISTIC CURVES

The circuits in figure 5-6 illustrate the biasing of PNP and NPN common base (CB) configurations. The *base* is the *common* terminal between the emitter input terminal and the collector output terminal. The parameters V_{EB}, I_E, V_{CB}, and I_C can be measured at the points shown in the circuits of figure 5-6. The emitter current, I_E, is the input current and the collector current, I_C, is the output current. The input voltage is V_{EB} and the output voltage is V_{CB}. To study the behavior of the CB transistor configuration, graphs are made of the relationship between I_E and V_{EB}, called the *input characteristic,* and the relationship between I_C and V_{CB}, called the *output characteristic.*

The CB input characteristics for an NPN transistor are determined by varying and recording the input voltage V_{EB} and the input current I_E for different values of the collector-base voltage V_{CB}. For a given value of V_{CB}, the graph of I_E versus V_{EB} is essentially the same as the graph for a *p-n* junction diode (that is, an emitter-base junction diode), as shown in figure 5-7.

Increases in V_{CB} cause the emitter-base junction to be a slightly better diode. That is, I_E becomes slightly larger for a given V_{EB} as V_{CB} increases. As shown in figure 5-7, the variation of output voltage V_{CB} has little effect upon the voltage across the emitter-base junction. In actual practice, manufacturers seldom provide input characteristic curves for transistors, but rather specify a value of V_{BE} for a given input current. The approximate voltages for V_{BE} as given in Table 5-1 are used in the remainder of this unit.

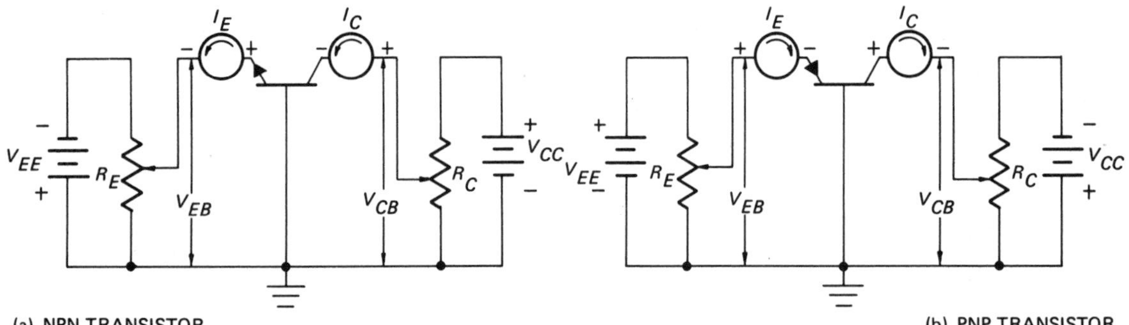

FIG. 5-6 CIRCUITS USED TO OBTAIN THE CB CHARACTERISTICS

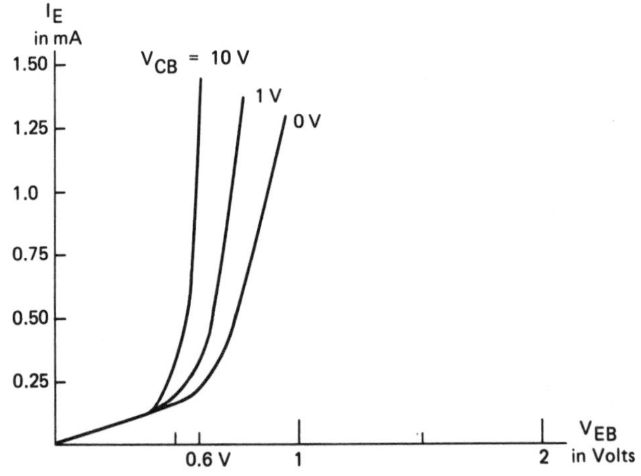

FIG. 5-7 COMMON-BASE INPUT CHARACTERISTICS FOR A TYPICAL NPN SILICON TRANSISTOR

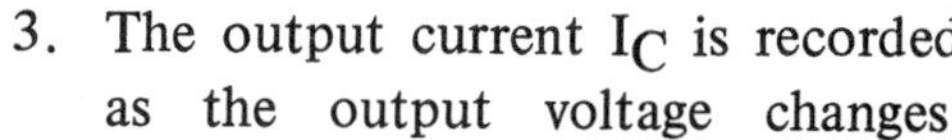

TABLE 5-1

Transistor type	V_{BE} in volts
Germanium (discrete)	0.2
Silicon (discrete)	0.6
Silicon (integrated circuit)	0.7

The most useful characteristic curves for a transistor are the *collector* or *output* characteristics. A typical set of collector characteristics for a transistor in the CB configuration is shown in figure 5-8. The curves for a PNP transistor are given in figure 5-8a; figure 5-8b gives the curves for an NPN transistor.

To determine the CB collector characteristics for an NPN transistor, the following steps are performed using the circuit of figure 5-6a.

1. The input current I_E is adjusted to a particular value, recorded, and held constant.
2. The output voltage V_{CB} is varied and recorded.
3. The output current I_C is recorded as the output voltage changes.
4. Steps 1, 2 and 3 are repeated for different values of the input current, I_E.

The resulting graph of I_C versus V_{CB} for the different values of I_E is shown in figure 5-8b. The dc and ac current gain of a transistor in the CB configuration can be determined from these characteristic curves using Eq. 5.4 and Eq. 5.5.

PROBLEM 1.

A PNP transistor has the characteristic curves shown in figure 5-8a. Determine the dc current gain of the transistor for a collector base voltage of -2.5 V and I_E of 15 mA. Find the base current for these conditions neglecting I_{CBO}.

$$\alpha = \left.\frac{I_C}{I_E}\right|_{V_{CB} = -2.5\ V} \qquad \text{Eq. 5.4}$$

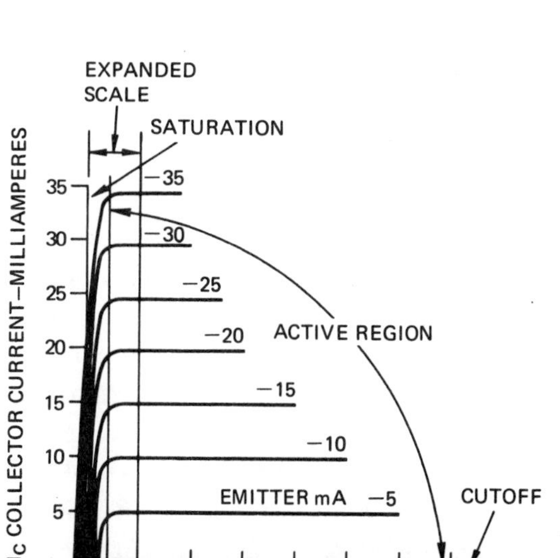

(a) PNP (GERMANIUM)

(b) NPN (SILICON)

FIG. 5-8 TYPICAL COLLECTOR CHARACTERISTICS FOR PNP AND NPN TRANSISTORS IN THE COMMON-BASE CONFIGURATION.

(Courtesy Institute of Electrical and Electronics Engineers, Inc.)

$$= \frac{14.6 \text{ mA}}{15.0 \text{ mA}} = \mathbf{0.972}$$

$$I_B = (1 - \alpha)\ I_E - I_{CBO} \quad \text{Eq. 5.6}$$

$$(1 - 0.972)\ 15 \text{ mA} - 0$$

$$= 0.028 \times 15 \text{ mA} = 0.43 \text{ mA}$$

$$= \mathbf{430\ \mu A}$$

PROBLEM 2.

A PNP transistor has the characteristic curves shown in figure 5-8a. Determine the ac current gain at a collector base voltage of -5 V when I_E changes from 20 mA to 25 mA. Find the base current change for these conditions, neglecting I_{CBO}.

$$\alpha_{ac} = \frac{\Delta I_C}{\Delta I_E} \quad V_{CE} = -5 \text{ V} \quad \text{Eq. 5.5}$$

$$= \frac{(24.3 - 19.6) \text{ mA}}{(25.0 - 20.0) \text{ mA}} = \mathbf{0.94}$$

$$\Delta I_B = (1 - \alpha_{ac})\ \Delta I_E - I_{CBO} \quad \textbf{Eq. 5.6}$$

$$= (1 - 0.94)\ (25 - 20) \text{ mA} - 0$$

$$= 0.06 \times 5 \text{ mA} = 0.3 \text{ mA}$$

$$= \mathbf{300\ \mu A}$$

Since the value for α in problem 2 is close to 1, typical common base collector characteristic curves can be drawn without actually measuring the necessary quantities. Carefully examine the curves shown in figure 5-8. Note that there is little change in the collector current as the collector-base voltage increases. After an initial change, the collector current is approximately equal to the emitter current. Since the common base collector characteristics are simple to predict, transistor manufacturers seldom provide these curves.

The curves in figure 5-8 also show that the collector current is not zero when I_E is zero. When the emitter current is zero, the current flowing in the collector circuit is the *leakage current* I_{CBO} or I_{CO}. Transistor specification sheets often refer to I_{CBO} as the collector cutoff current. The region of operation below the value of the leakage current is called the *cutoff region*.

To operate in the cutoff region, the emitter-base junction of the transistor must have zero bias or must be reverse biased. As a result, the only current flowing in the collector circuit is the minute amount of leakage current due to the reverse biased collector-base junction. The magnitude of the leakage current is expressed in nanoamperes (nA) in silicon and in microamperes (μA) in germanium. The transistor output voltage V_{CB} will be approximately equal to the supply voltage, since the voltage drop across the collector resistor is small.

When the emitter-base junction is forward biased, emitter current will flow. Most of the emitter current will flow as collector current. The region of operation where the collector current is approximately equal to the emitter current is called the *active region* and is shown in figure 5-8. The current equations for the CB active region were presented earlier in this unit.

If the emitter current is increased, a point is reached beyond which the collector current will no longer increase. The term *saturation* is applied to this condition. In the saturation regions shown in figure 5-8, note that the voltage across the collector-base junction of the transistor is nearly equal to zero. In this region, the collector-base junction is actually forward biased. For example, in saturation the PNP transistor collector voltage V_{CB} is slightly positive. Since the collector-base voltage is small, the collector current depends upon the supply voltage, V_{CC}, and the collector resistor, R_C, as shown in figure 5-6. An approximate value for the collector current in the saturation region is expressed by the equation:

$$I_{CS} = \frac{V_{CC}}{R_C} \qquad \text{Eq. 5.7}$$

where V_{CC} = collector supply voltage

R_C = resistor in series with the collector

I_{CS} = collector saturation current

For both PNP and NPN transistors, the three regions of operation are the *active*, *saturation*, and *cutoff* regions. Figure 5-9

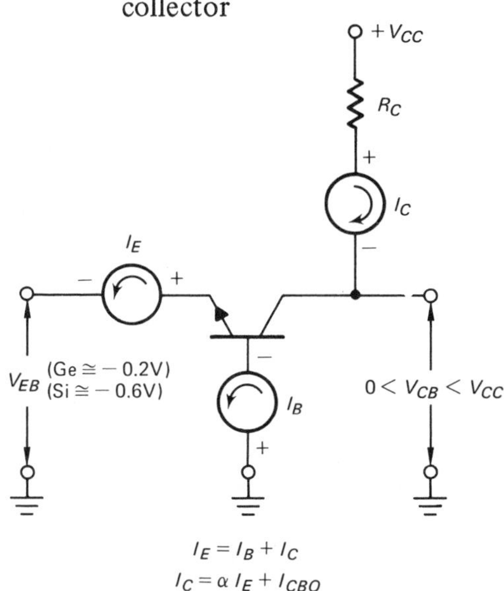

ACTIVE REGION OPERATION WITH EMITTER-BASE JUNCTION *FORWARD* BIASED AND COLLECTOR-BASE JUNCTION *REVERSED* BIASED.

(a)

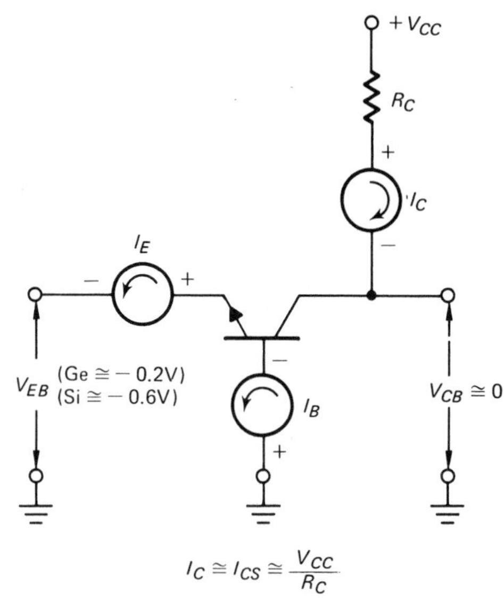

SATURATION REGION OPERATION WITH EMITTER-BASE JUNCTION *FORWARD* BIASED AND COLLECTOR-BASE JUNCTION NEARLY EQUAL TO *ZERO* VOLTS.

(b)

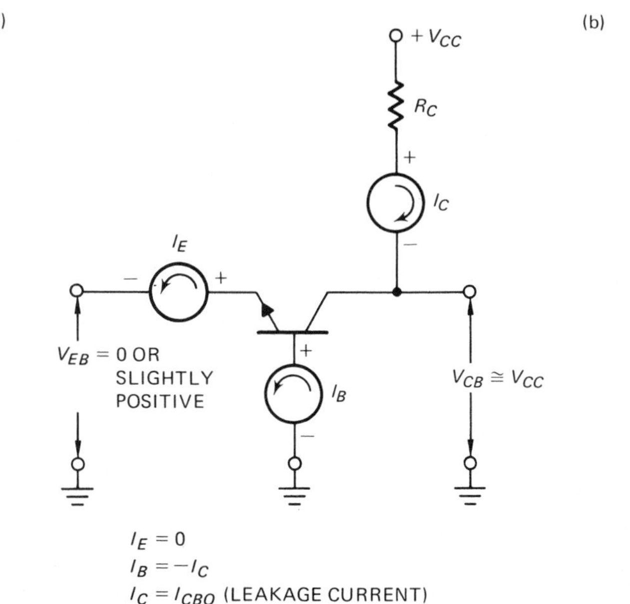

CUTOFF REGION OPERATION WITH EMITTER-BASE JUNCTION *REVERSE* BIASED AND COLLECTOR-BASE JUNCTION *REVERSE* BIASED.

(c)

FIG. 5-9 THREE REGIONS OF OPERATION FOR A COMMON-BASE NPN TRANSISTOR

shows the CB biasing required for an NPN transistor and the voltages that exist across the transistor in each region.

Note the battery symbol for the collector supply voltage V_{CC} in figure 5-9. As shown, the positive side of the battery is connected to the collector resistor. The negative side of the battery, however, is understood to be connected to ground. Refer to figure 5-10 for an interpretation of the battery symbol.

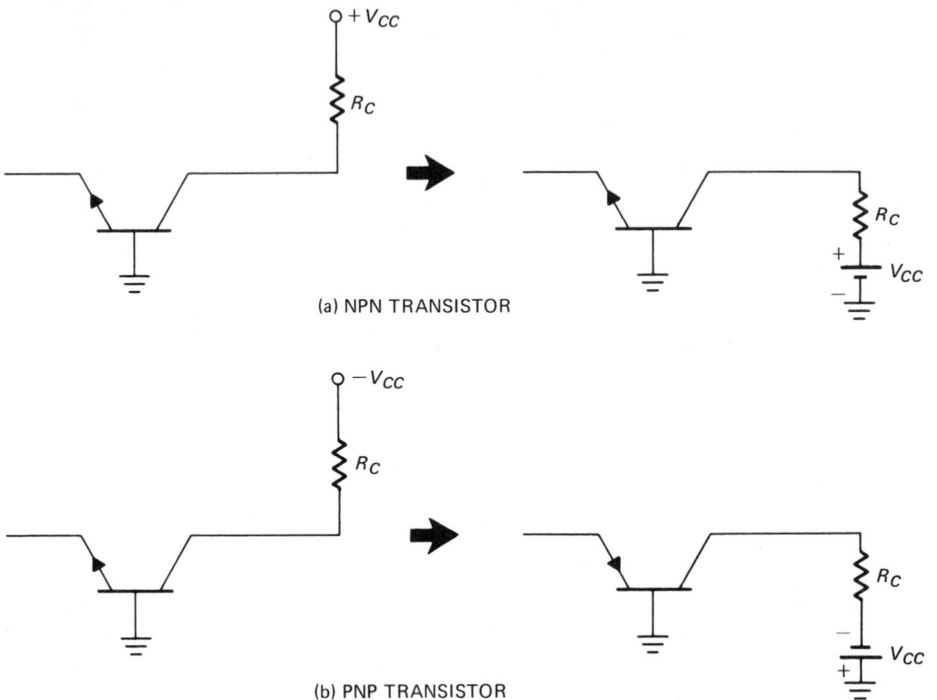

FIG. 5-10 SYMBOLIZING THE BATTERY FOR TRANSISTOR AMPLIFIER BIASING

LABORATORY EXERCISE 5-1: TRANSISTOR FAMILIARIZATION

PURPOSE

- To observe the operation of an NPN and a PNP transistor in the three regions of operation.
- To determine the relation of collector and emitter current in the CB configuration.
- To measure the leakage current I_{CBO}.
- To compare actual laboratory data with theoretical calculations for collector and emitter currents.

MATERIALS

1 NPN transistor, 2N4425 or any small signal transistor
1 PNP transistor, 2N525 or any small signal transistor
1 Dc power supply, or 1.5 V battery
1 Dc power supply, 6 V
2 Milliammeters, multirange, or 20,000 ohm per volt VOM

1 VTVM, or solid-state voltmeter
1 Resistor, 68Ω, 1/2 W
1 Resistor, 1 kΩ, 1/2 W
1 Potentiometer, linear, 2W
2 Switches, single pole, single throw (SPST)

PROCEDURE

A. 1. Complete the diagrams in figure 5-11 to show that the transistor is biased in the *active* region. Label the transistor terminals E, B, and C. Show the correct polarity on the ammeters and insert the power supplies.

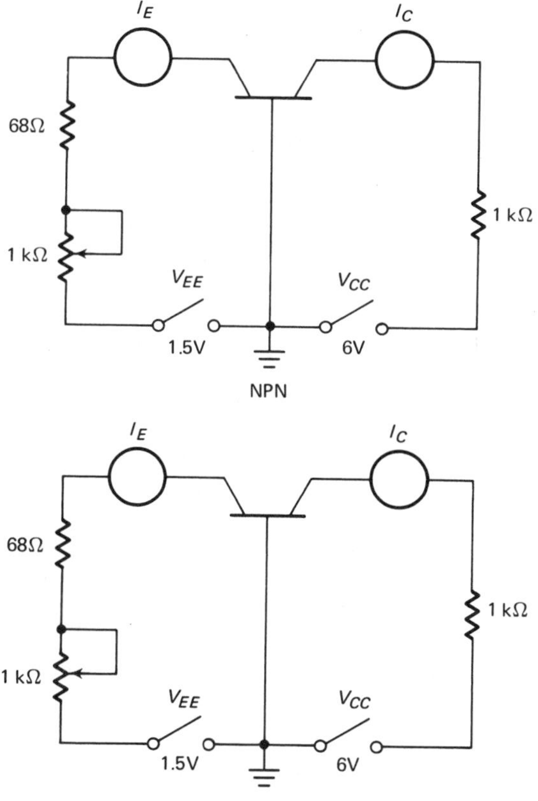

FIG. 5-11 LABORATORY CIRCUIT FOR TRANSISTOR FAMILIARIZATION

2. Calculate the maximum collector current that can flow.

$$I_{CS} = \frac{V_{CC}}{R_C}$$

3. Assuming that the transistor is silicon ($V_{BE} = 0.6$ V), calculate the maximum and minimum emitter current that can flow.

$$I_{E\,(MAX)}\ (\text{Pot} = 0) = \frac{V_{EE} - V_{BE}}{R_E} = \frac{(1.5 - 0.6)\ \text{V}}{68\ \Omega} = 13.2\ \text{mA}$$

$$I_{E\,(MIN)}\ (\text{Pot} = 1\ \text{k}\Omega) = \frac{(1.5 - 0.6)\ \text{V}}{1068\ \Omega} = 0.844\ \text{mA}$$

4. Assuming that the transistor is germanium ($V_{BE} = 0.2$ V), calculate the maximum and minimum emitter current that can flow.

   $I_{E\,(MAX)} = ?$

   $I_{E(MIN)} = ?$

5. If the collector current is equal to 3 mA, calculate the collector to base voltage.
   $V_{CB} = V_{CC} - I_C\ R_C = 6\ \text{V} - (3\ \text{mA})\ (1\ \text{k}\Omega) = 6\ \text{V} - 3\ \text{V} = 3\ \text{V}$

6. If the switch in the emitter base circuit is open and the switch in the collector-base circuit is closed, predict the emitter and collector currents and the collector-base voltage.

   $I_E = ?$

   $I_C = ?$

   $V_{CB} = ?$

B. 1. Connect the NPN transistor circuit as shown in figure 5-11.
2. Set the potentiometer for the maximum resistance.
3. Leave the switches open until readings are to be taken.

C. 1. Close the switches and measure V_{EE} and V_{CC}.
2. Set the potentiometer in the emitter-base circuit so that the emitter current equals 3 mA (the collector current is approximately the same). The transistor is now in the active region.
3. Take all of the data required for step 2 in Table 5-2. Be sure to show the voltage polarity.

TABLE 5-2 TRANSISTOR FAMILIARIZATION

	NPN Transistor					PNP Transistor				
Region of Operation	Step	Emitter-Base Circuit		Collector-Base Circuit		Step	Emitter-Base Circuit		Collector-Base Circuit	
		I_E in mA	V_{EB} in V	I_C in mA	V_{CB} in V		I_E in mA	V_{EB} in V	I_C in mA	V_{CB} in V
Active	2					2				
E-B Forward Biased (No V_{CC})	3									
Saturation	4					4				
Cutoff	5					5				
I_{CBO}	6									

D. To observe the need for a collector-base supply, set V_{CC} to zero (or remove V_{CC} and replace it by a short). Take the data required for step 3 in Table 5-2.

E. 1. Set V_{CC} to 6 V. Using the potentiometer, decrease the resistance in the emitter-base circuit. Observe the milliammeters as this is done. At first both I_E and I_C will increase. A point is reached, however, at which I_C no longer increases. This point is the saturation current I_{CS}. Continue to turn the potentiometer until it is set at minimum resistance.

2. Record all readings for step 4 in Table 5-2 (Note the polarity on the V_{CB} reading).

F. 1. Adjust the potentiometer until the transistor is operating in the active region (I_E = 3 mA).

2. Reverse the emitter-base battery or power supply. The emitter-base junction of the transistor is now reverse biased.

3. Record all readings for step F in Table 5-2. (A microampere scale is required for the I_C reading)

G. 1. Adjust the circuit so that the transistor is again operating in the active region (I_E = 3 mA). Now open the switch in the emitter-base circuit.

2. Record all readings for step G in Table 5-2.

H. 1. Connect the PNP transistor circuit as shown in figure 5-11, with the potentiometer set for maximum resistance.

2. Repeat steps C, E, and F for the PNP transistor circuit and record all data in Table 5-2.

DISCUSSION QUESTIONS

1. Discuss the operation of the transistors in each of the regions of operation. The discussion should include the biasing required for the emitter-base and collector-base junctions, and should provide typical values for the collector current and the collector-base voltage for each region. How do the readings for the NPN and PNP transistors differ?
2. In step C.2., why should the collector current be approximately equal to the emitter current? Can the collector current exceed the emitter current?
3. From the data taken in step C. for both the NPN and PNP transistors, calculate α_{dc}.
4. In step E., state why I_C does not follow I_E all the way as the resistance in the emitter-base circuit is decreased.
5. What is I_{CEO}? In which steps was I_{CBO} measured?

6. Compare the theoretical calculations in step A to the actual readings in Table 5-2. Discuss these comparisons.

LABORATORY EXERCISE 5-2: CB COLLECTOR CHARACTERISTICS

PURPOSE

- To obtain the CB collector characteristics for an NPN transistor.
- To plot the CB collector characteristics and determine the current gain of the CB configuration from the plot.

MATERIALS

1 NPN transistor, 2N4425 or 2N3405 or equivalent
1 Battery, 1.5 V, or dc power supply
1 Dc power supply, 12 V
2 Multirange milliameters, or 20,000 ohm per volt VOMs
1 VTVM, or solid-state voltmeter
1 Linear potentiometer, 1 kΩ, 2 W
1 Linear potentiometer, 5 kΩ, 2 W
2 Switches, SPST

PROCEDURE

A. 1. From figure 5-6, find the correct circuit to obtain the CB characteristics for an NPN transistor. Redraw the circuit and insert a 12-V power supply for V_{CC}, a 1.5-V power supply for V_{EE}, a 1-kΩ potentiometer for R_C, a 5-kΩ potentiometer for R_E, and a switch by each power supply.

2. Find the maximum collector power dissipation rating for the transistor, P_C = ? mW. For this experiment, limit the actual P_C to two-thirds the maximum P_C. Using the experimental limit P_C, calculate the collector current for each value of the collector-base voltage shown in Table 5-3.

$$I_C = \frac{P_C}{V_{CB}}$$

TABLE 5-3

V_{CB} (volts)	0.5	1	2	3	4	5	6	8	10	12
I_C (mA)										

3. Read the balance of the procedure to determine the CB collector characteristics given in the unit.

B. 1. Connect the circuit drawn in step A.1.

2. Set the potentiometers so that voltage is not appplied across the emitter-base and collector-base circuits.

3. Leave the switches open until readings are to be taken.

C. 1. To check for the proper operation of the circuit, close the switches and set the potentiometer R_E so that an emitter current of approximately 1 mA is flowing.

2. Increase the collector-base voltage by adjusting the potentiometer R_C. The collector current should increase up to the emitter current and then stay constant as V_{CB} is increased.

3. If the circuit performs as described, then the CB characteristics can be run. If the circuit does not perform properly, consult the instructor.

D. 1. Set the potentiometer R_E so that V_{EB} equals 0 volts. At this time I_E is equal to 0 mA. (I_E may be set equal to zero by removing both V_{EE} and R_E and shorting the emitter and base leads of the transistor together).

2. Vary V_{CB} through the range shown in Table 5-4.

3. Record the collector current at each value of V_{CB} in Table 5-4. (A microampere scale is required for the I_C reading.)

TABLE 5-4 CB COLLECTOR CHARACTERISTICS

I_E	I_C (mA) measured at the following values of V_{CB}										
in mA	0V	0.5V	1.0V	2V	3V	4V	5V	6V	7V	9V	11V
0											
1											
2											
3											
4											
5											
6											
8											
10											
15											
20											

E. 1. Adjust the potentiometer R_C so that V_{CB} is equal to 0 V. Then adjust potentiometer R_E for the next value of I_E as shown in Table 5-4.

2. Vary V_{CB} through the range shown in Table 5-4 and record the collector current. The emitter current must remain constant as V_{CB} is varied through its range; it may be necessary to adjust potentiometer R_E as potentiometer R_C is varied to obtain V_{CB}.

F. Repeat step E. for each value of emitter current shown in Table 5-4. Do not exceed the collector currents calculated at each value of V_{CB} in step A.2.

G. 1. From the data in Table 5-4, plot the CB collector characteristic curves on graph paper.

2. Plot the collector power dissipation curve calculated in step A.2, Table 5-3.

H. If a transistor curve tracer is available, connect the transistor in the CB configuration. Compare these curves with those plotted for step G.

DISCUSSION QUESTIONS

1. When V_{CB} is equal to zero volts, I_C is almost equal to I_E. For all values of I_E, state why I_C is not approximately equal to 0 mA when V_{CB} is equal to 0. Refer to figure 5-8 and discuss this situation.
2. As the collector-base voltage is varied through its range with a fixed value of emitter current, how does the collector current vary? Which has a greater effect on collector current, the collector-base voltage or the emitter current?
3. From the curves drawn in step G of the procedure, calculate α_{dc} at each of the following locations:

   $V_{CB} = 5$ V, $I_E = 1$ mA .

   $V_{CB} = 5$ V, $I_E = 10$ mA

   Using these current gains, find the base current for each value of I_E (Neglect I_{CBO}).
4. Using the curves drawn in step G of the procedure, determine the ac current gain of the transistor at a collector-base voltage of 5 V and an I_E change from 8 mA to 10 mA.
5. What quantity is measured in step D?
6. What is the current gain of a transistor in the CB configuration?

EXTENDED STUDY TOPICS

1. Describe the current flow in a PNP transistor.
2. Does leakage current always flow in a transistor? Explain.
3. Why is the base region lightly doped?
4. Determine the dc current gain for the transistor characteristics in figure 5-8a for a collector base voltage of -3.0 V and $I_E = 10$ mA. Neglect I_{CBO}.
5. Find the base current for Topic 4.
6. Determine the ac current gain for the transistor characteristics in figure 5-8b at a collector base voltage of 5 V and an I_E change of -15 mA to-10 mA.
7. Find the base current change for Topic 6. Neglect I_{CBO}.
8. Define the active region.
9. Define the saturation region.
10. Define the cutoff region.

Common-base configuration

OBJECTIVES

After studying this unit, the student will be able to discuss and demonstrate an understanding of the basic principles of:

- Simple common-base amplifiers
- Transistor current and voltage calculations
- Common-base input and output resistance and voltage gain calculations

SIMPLE COMMON-BASE AMPLIFIERS

Typical common-base amplifier circuits for PNP and NPN transistors biased in the active region are shown in figure 6-1. The input dc supply is V_{EE} and the collector supply is V_{CC}.

For the basic CB amplifier circuits for PNP and NPN transistors in figure 6-2, an increase in the input dc power supply V_{EE} will result in an increase in the emitter current. Recall that the emitter-base voltage drop across a transistor remains essentially constant. As a result, most of the increase in V_{EE} must be dropped across the emitter resistor R_E when the emitter current increases. As long as the transistor operation remains in the active region, the collector current also increases since $I_C = \alpha I_E + I_{CBO}$. Due to the collector current increase, the voltage drop across the collector resistor will increase with a resulting decrease in the collector-base voltage V_{CB}.

To summarize, for the PNP transistor in figure 6-2, as the emitter becomes more positive, V_{CB} becomes less negative (or more positive) than it was before the change in emitter current. Similarly, for the NPN transistor in figure 6-2, as the emitter becomes more negative, V_{CB} becomes less positive (or more negative) than it was before the change.

If the input power supply V_{EE} is decreased, the emitter current also decreases, causing a reduction in the collector current and the voltage drop across the collector resistor. As a result, the collector-base voltage V_{CB} increases.

In the active region of transistor operation, if V_{EE} is held constant and the value of V_{CC} is increased, thus increasing the voltage

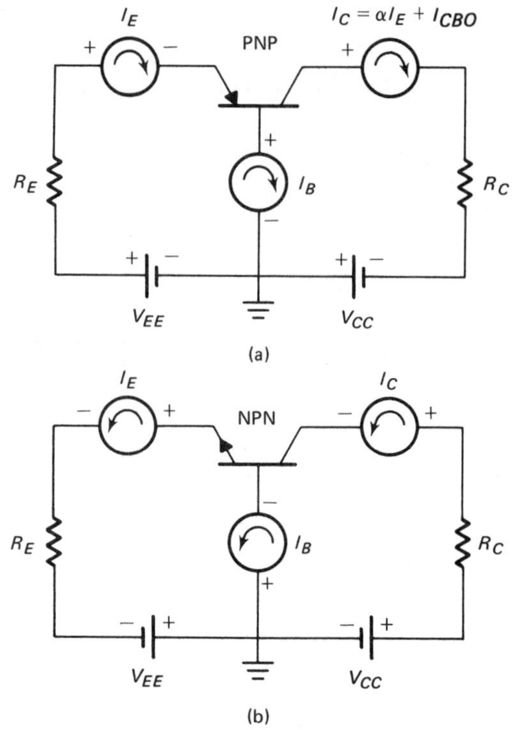

FIG. 6-1 COMMON-BASE CONFIGURATION FOR ACTIVE REGION

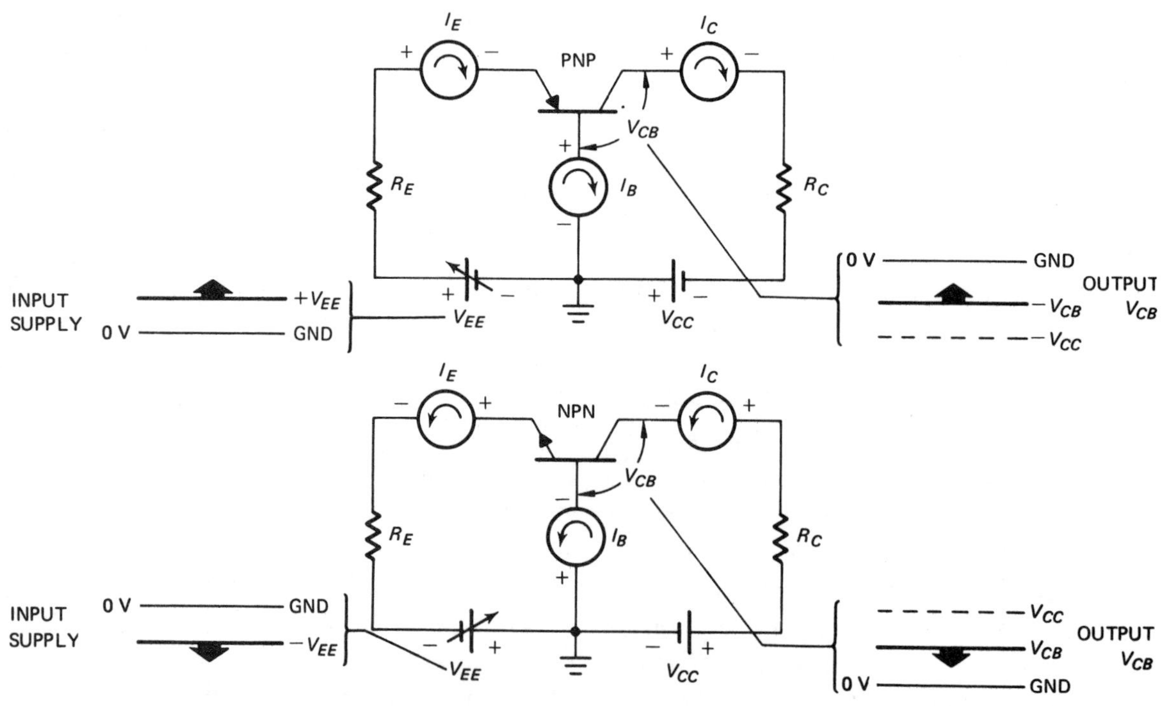

FIG. 6-2 COMMON-BASE AMPLIFIER CIRCUIT SHOWING EFFECT OF CHANGES OF INPUT SUPPLY UPON OUTPUT VOLTAGE

across the collector-base junction of the transistor, the collector current shows little change, if any. Refer to the common base characteristic curves given in figure 6-3. These curves indicate that the collector-base voltage has little control over the collector current after the initial reverse bias is applied.

As V_{EE} is increased, the collector current increases to the point where it is approximately equal to V_{CC}/R_C. Although V_{EE} is increased beyond this point, the collector current cannot increase. This condition is known as *saturation*.

If the polarity of V_{EE} is reversed so that the emitter-base junction is reverse biased, then the resistor is cut off. In this condition, no emitter current will flow; the only collector current flowing is the leakage current. At cutoff, the collector-base voltage V_{CB} is approximately equal to the supply voltage V_{CC}.

The preceding paragraphs illustrate the need for resistors R_E and R_C in the common-base amplifier circuit. Resistor R_E limits the emitter-current. As the dc supply V_{EE} is increased in magnitude, the voltage is dropped across R_E. The size of the resistor R_E depends upon the required emitter current and the emitter-base power supply.

The output of the common-base circuit is taken across the collector-base junction of the transistor. When the input voltage is changed, it is expected that the output voltage will change also. If the collector circuit does not contain a resistor, then the voltage across the collector-base junction will equal the collector supply voltage V_{CC}. With resistor R_C in the collector circuit, then as the collector current changes, the voltage drop across R_C also changes, resulting in a change in output voltage.

PROBLEM 1.

A common-base amplifier circuit contains a silicon NPN transistor as shown in figure 6-4. Find a. the emitter current, I_E, b. the collector current, I_C, c. the collector-base voltage, V_{CB}, and d. define the region of operation.

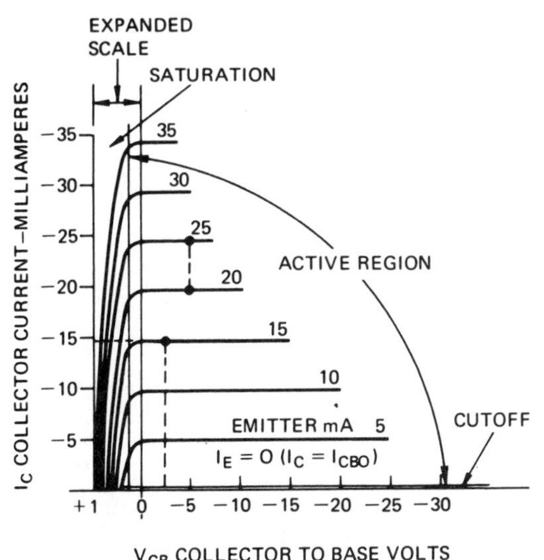

(a) PNP (GERMANIUM)

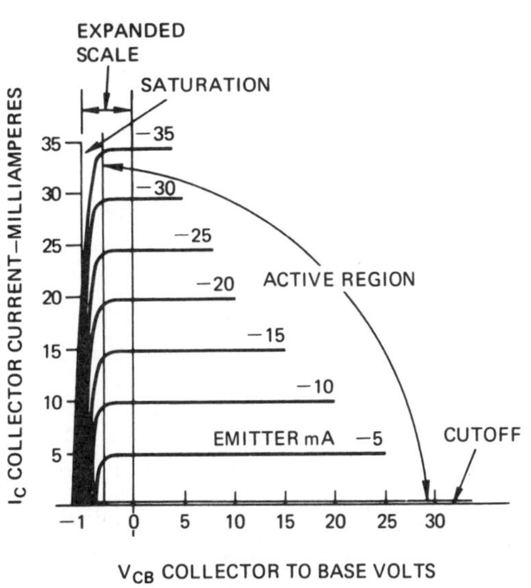

(b) NPN (SILICON)

FIG. 6-3 TYPICAL COLLECTOR CHARACTERISTICS FOR PNP AND NPN TRANSISTORS IN THE COMMON-BASE CONFIGURATION.

(Courtesy Institute of Electrical and Electronics Engineers, Inc.)

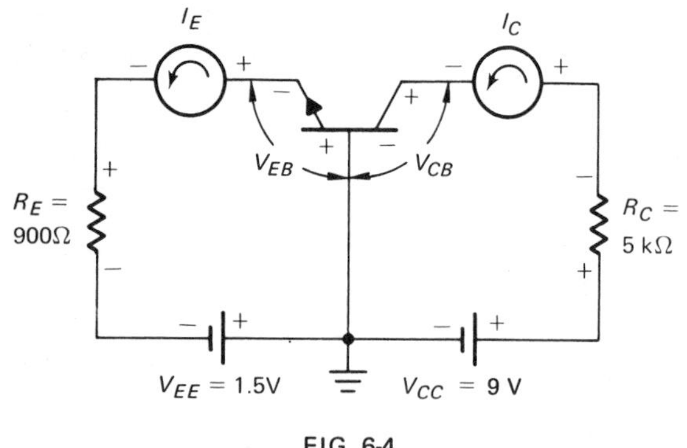

FIG. 6-4

a. Since a silicon transistor is used, the voltage across the emitter-base junction is 0.6 V. The voltage across the 900-ohm emitter resistor can be found by using Kirchhoff's voltage law.

$V_E = 1.5\text{ V} - V_{EB} = 1.5\text{ V} - 0.6\text{ V}$
$= 0.9\text{ V}$

I_E can be found using Ohm's Law:

$$I_E = \frac{0.9\text{ V}}{900\ \Omega} = \frac{0.9\text{ V}}{0.9\text{ k}\ \Omega} = \mathbf{1.0\ mA}$$

b. The current gain α is not given. Since α is always near unity, *assume* that $\alpha = 1$. The leakage current I_{CBO} is not given; therefore, assume that the leakage current is zero. To determine the collector current:

$I_C = \alpha I_E + I_{CBO} = 1 \times 1\text{ mA} + 0.0$
$I_C = 1\text{ mA}$

c. Use Kirchhoff's voltage law in the collector-base circuit to find V_{CB}.

$V_{CB} = V_{CC} - I_C R_C$
$= 9.0\text{ V} - 1.0\text{ mA} \times 5.0\text{ k}\Omega$
$= 9.0\text{ V} - 5.0\text{ V} = \mathbf{4.0\ V}$

d. Since the emitter-base junction is forward biased, it must be in either the active or the saturation region. To be saturated, V_{CB} must be near zero and the collector-base junction must be forward biased. Since $V_{CB} = +\ 4.0$ V (step c.), the collector-base junction is reverse biased; therefore, the region of operation is the active region.

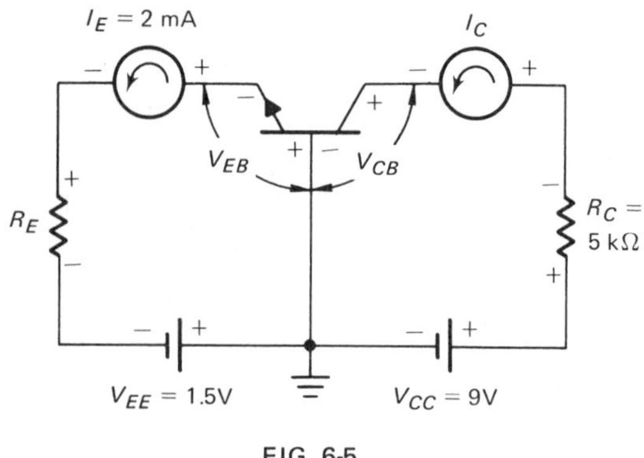

FIG. 6-5

PROBLEM 2

A common-base amplifier circuit contains a silicon NPN transistor as shown in figure 6-5. Find a. the collector current, I_C, and b. define the region of operation.

a. Assuming that $\alpha = 1$, then I_C
$= I_E = 2\text{ mA}$

b. Calculate the collector-base voltage V_{CB}. Applying Kirchhoff's voltage law to the collector circuit yields:

$V_{CB} = V_{CC} - I_C R_C$
$= 9\text{ V} - 2\text{ mA} \times 5\text{ k}\Omega = 9\text{ V} - 10\text{ V}$
$= -1\text{ V}$

$V_{CB} = -1$ V indicates that the collector-base junction is forward biased. As a result, the transistor is *saturated* and the value of the collector current must be recalculated.

$$I_{CS} = \frac{V_{CC}}{R_C}$$

$$= \frac{9\text{ V}}{5\text{ k}\Omega} = \mathbf{1.8\ mA}$$

Using the transistor characteristics shown in figure 6-3a, find α, I_C, and I_B if $I_E = 36$ mA, $V_{CB} = -5$ V, and $I_{CBO} = 0$. (R6-1)

Determine I_C and I_E if $I_B = 0.11$ mA, $\alpha = 0.93$, and $I_{CBO} = 10\ \mu A$. (R6-2)

Determine the ac current gain for V_{CB} = -5 V and an I_E change of 25 mA to 20 mA using figure 6-3a. Find the base current change for these conditions if I_{CBO} = 10 μ A. (R6-3)

COMMON-BASE INPUT AND OUTPUT RESISTANCE AND VOLTAGE GAIN

In units 5 and 6, the input and output characteristic curves for the CB transistor have been presented and analyzed. Certain parameters for the CB transistor can be obtained from these curves. One parameter is the current gain. Two other parameters which can be found for the CB transistor are the *input resistance,* r_{ib}, and the *output resistance*, r_{ob}. These two parameters are expressed mathematically as follows:

$$r_{ib} = \frac{\Delta V_{EB}}{\Delta I_E} \qquad \text{Eq. 6.1}$$

where ΔV_{EB} = change in emitter-base voltage

ΔI_E = change in emitter current

r_{ib} = input resistance of a CB transistor

$$r_{ob} = \frac{\Delta V_{CB}}{\Delta I_C} \qquad \text{Eq. 6.2}$$

where ΔV_{CB} = change in collector-base voltage

ΔI_C = change in collector current

r_{ob} = output resistance of a CB transistor

The following problems illustrate and interpret the approximate values for r_{ib} and r_{ob}.

PROBLEM 3

A transistor experiences a change in V_{EB} of 2 mV when the emitter current changes by 0.1 mA. Find the resistance r_{ib}.

$$r_{ib} = \frac{\Delta V_{EB}}{\Delta I_E} = \frac{2.0\text{ mV}}{0.1\text{ mA}} = \textbf{20 ohms}$$

This problem shows that the input resistance of the CB transistor is small since the emitter-base junction is forward biased. Typically, the CB input resistance ranges between 10 and 1000 ohms.

PROBLEM 4

A transistor experiences a change in V_{CB} of 10 V when the collector current changes by 10 μA. Find the resistance r_{ob}.

$$r_{ob} = \frac{\Delta V_{CB}}{\Delta I_C} = \frac{10\text{ V}}{10\ \mu\text{A}} = 1\text{ M}\Omega$$

This problem shows that the output resistance of the CB transistor is large since the collector-base junction is reverse biased. Typically, the output resistance is greater than 1 M Ω.

Figure 6-6 shows an NPN CB transistor amplifier which operates in the active region. The emitter is negative with respect to the base, and the collector is positive with respect to the base. (Recall that the PNP transistor has the opposite polarity.) The circuits of figure 6-6 can be used to illustrate the various characteristics of the CB amplifier.

In figure 6-6b, the transistor input resistance, r_{ib}, is in parallel with the emitter resistor R_E. This parallel combination of resistance is the *amplifier input resistance.* This is the resistance seen by a voltage source connected to the input of the circuit.

The value of r_{ib} normally is much smaller than R_E. As a result, the amplifier input resistance is equal to r_{ib} and, as stated above, is the resistance seen by the voltage source. The following problem illustrates this point.

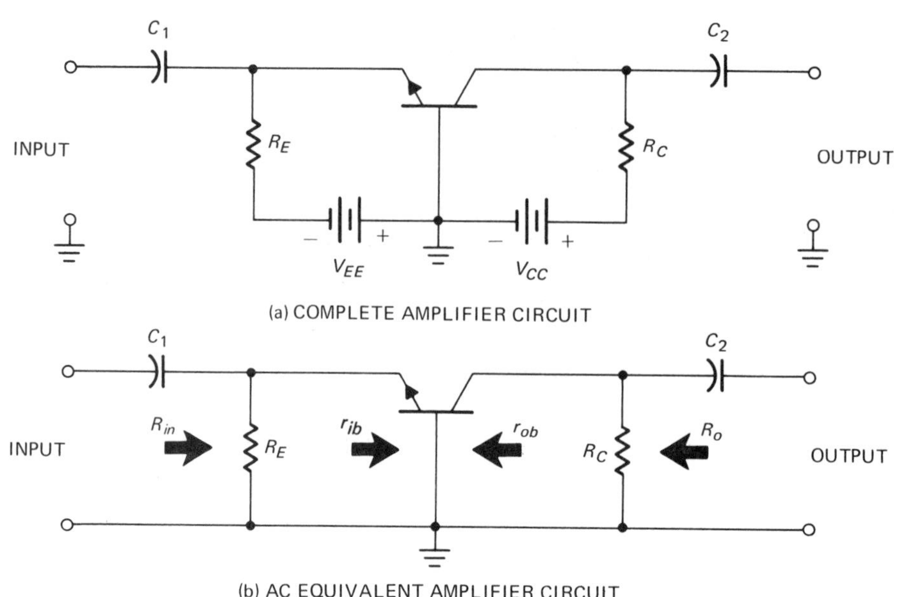

FIG. 6-6 TYPICAL CB AMPLIFIER CIRCUIT EMPLOYING NPN TRANSISTOR

PROBLEM 5.

A CB amplifier has R_E = 900 ohms, and the transistor has an r_{ib} of 20 ohms. Find the amplifier input resistance.

R_{in} is equal to the parallel combination of r_{ib} and R_E. Mathematically, this is expressed as:

$$R_{in} = r_{ib} \,||\, R_E = 20\ \Omega \,||\, 900\ \Omega$$

$$\text{or} \quad R_{in} = \frac{20 \times 900}{20 + 900} = \mathbf{19.6\ \Omega}$$

In figure 6-6b, the transistor output resistance, r_{ob}, is in parallel with the collector resistance R_C. This parallel combination of resistance is the *amplifier output resistance.* This is the resistance seen by another transistor amplifier if it is connected to the output of the circuit. The value of R_C normally is much smaller than r_{ob}; thus, the amplifier output resistance is equal to R_C.

PROBLEM 6.

A CB amplifier circuit has R_C = 5000 ohms, and the transistor has an r_{ob} of 1 M Ω. Find the amplifier output resistance.

$$R_o = r_{ob} \,||\, R_C = 1\ \text{M}\Omega \,||\, 5000\ \Omega$$
$$= \mathbf{5000\ \Omega}$$

The voltage gain, A_v, for any electronic system is defined as the ratio of the output voltage change to the input voltage change. Eq. 6.3 expresses this relationship.

$$A_v = \frac{\Delta V_{out}}{\Delta V_{in}} \qquad \textbf{Eq. 6.3}$$

where Δ = a change in a physical quantity

ΔV_{out} = the change in output voltage

ΔV_{in} = the change in input voltage

A_v = voltage gain

Eq. 6.3 can be used to find the voltage gain for a CB amplifier by making the following observations in the circuit of figure 6-6.

$$\Delta V_{out} = \Delta I_C \times R_C$$

$$\Delta V_{in} = \Delta I_{in} \times R_{in}$$

Substituting the expressions for ΔV_{out} and ΔV_{in} in Eq. 6.3 yields:

$$A_v = \frac{\Delta I_c \times R_c}{\Delta I_{in} \times R_{in}}$$

$$\Delta I_E = \frac{R_E}{R_E + r_{ib}} \times \Delta I_{in}$$

Therefore,

$$\Delta I_{in} = \Delta I_E \left(\frac{R_E + r_{ib}}{R_E}\right)$$

$$R_{in} = \frac{r_{ib}\, R_E}{r_{ib} + R_E}$$

Substituting the expressions for ΔI_{in} and R_{in} in the voltage gain equation yields:

$$A_v = \frac{\Delta I_c R_c}{\left[\frac{R_E + r_{ib}}{R_E}\right]\Delta I_E\left[\frac{r_{ib} R_E}{r_{ib} + R_E}\right]}$$

and

$$A_v = \frac{\Delta I_c R_c}{\Delta I_E\, r_{ib}}$$

The ratio of the changing collector current to the changing emitter current $\left(\frac{\Delta I_c}{\Delta I_E}\right)$ in the equation above is equal to α_{ac}.

Thus, $A_v = \dfrac{\alpha R_C}{r_{ib}}$

Since $\alpha_{ac} \cong 1$, the voltage gain equation can be written as:

$$A_{v\,(CB)} = \frac{R_c}{r_{ib}} \qquad \text{Eq. 6.4}$$

Equation 6.4 is only an approximation of the CB voltage gain; however, this approximate is sufficient for the present analysis.

PROBLEM 7.

A CB amplifier circuit has the following parameters: R_E = 900 ohms and R_C = 5000 ohms. The circuit transistor has an r_{ib} = 20 Ω and an r_{ob} = 1 M Ω. Find the approximate voltage gain.

$$A_v \cong \frac{R_C}{R_{ib}} = \frac{5000\,\Omega}{20\,\Omega}$$

$$= 250$$

As shown in problem 7, the voltage gain for a CB transistor amplifier is high. Typically, the voltage gain varies between 50 and 500. As a result, the CB configuration makes an excellent voltage amplifier. Recall, however, that the current gain for a CB transistor is essentially α_{ac} (or h_{fb}) which is nearly unity.

A transistor has an r_{ib} = 30 Ω and α = 0.98, when V_{BE} = 0.3 V, I_E = 1 mA, and I_C = 0.96 mA. Determine the increase in I_E and I_C if V_{BE} increases to 0.31 V. (R6-4)

Find the voltage gain in (R6-4) if ΔV_{BC} = 15 V. (R6-5)

LABORATORY EXERCISE 6-1: COMMON-BASE AMPLIFIER

PURPOSE

- To predict the dc and ac performance of a common-base amplifier.
- To construct and then observe the operation of a common-base amplifier.
- To measure the dc operating point of a common-base amplifier.
- To measure the ac performance quantities, such as input and output resistance and voltage and current gain, for a common-base amplifier.

MATERIALS

1 NPN transistor, 2N4425, 2N3405 or equivalent
1 Battery, 1.5 V, or dc power supply
1 Dc power supply, 9 V

1 VTVM, or solid-state voltmeter
1 VTVM, ac (use an oscilloscope if none is available) ·
1 Signal generator
1 Oscilloscope
2 Electrolytic capacitors, 25 μ F, 50 V
1 Resistor, 100 Ω, ½ W
1 Resistor, 180 Ω, ½ W
1 Resistor, 1 k Ω, ½ W
1 Linear potentiometer, 1 k Ω, 2 W
2 Switches to connect the power supplies to the circuit, SPST

PROCEDURE

A. 1. Redraw the complete CB amplifier circuit given in figure 6-6a. Insert a 9-V power supply for V_{CC}, a 1.5-V power supply for V_{EE}, $C_1 = C_2 = 25\ \mu F$, $R_C = 1\ k\Omega$, $R_E = 180\ \Omega$, and a switch by each power supply.

2. List the results of the following calculations in Tables 6-1 and 6-2.

TABLE 6-1 Dc (biasing) Conditions

	V_{EE} in V	V_{EB} in V	I_E in mA	I_C in mA	V_{CB} in V
Theoretical Values					
Actual Test Results					

TABLE 6-2 Ac Performance Quantities

	r_{ib} in Ω	i_{in} in A	R_{in} in Ω	i_o in A	R_o in Ω	A_v (CB)	A_i (CB)
Theoretical Values		——		——			
Actual Test Results	——						

3. Assume that an NPN silicon transistor is used in this circuit. Calculate the emitter current.

$$V_{EE} = I_E R_E + V_{BE}$$

$$1.5\ V = 180\ I_E + 0.6\ V$$

$$I_E = ?$$

4. Assume that α is equal to 1. Calculate the collector current and the collector-base voltage.

$$I_C = \alpha I_E$$

$$V_{CC} = I_C R_C + V_{CB}$$

therefore,

$$V_{CB} = V_{CC} - I_C R_C$$

5. An approximation for the input resistance of the emitter-base junction of a transistor is:

$$r_{ib} \cong \frac{50 \text{ mV}}{I_E}$$

Find r_{ib} for the value of I_E calculated in step A.3.

6. Find the amplifier input resistance (R_{in}).

$$R_{in} = r_{ib} \,||\, R_E$$

7. Assuming that the output resistance of the transistor, r_{ob}, is very large, what is the amplifier output resistance, R_O?
8. Find the approximate voltage gain for this amplifier.

$$A_{v\,(CB)} = \frac{R_C}{r_{ib}}$$

9. Find the approximate current gain for this amplifier. Assume that most of the input ac current flows to the emitter of the transistor; that is, there is very little current lost in the 180-Ω resistor.

$$A_{i\,(CB)} \cong \alpha$$

B. Connect the circuit as drawn in step A.1.

C. 1. Check the operating point of the circuit; that is, measure the dc voltages V_{CB} and V_{EB}. The voltage from the collector to ground or the base of the transistor, V_{CB}, should be between 3 and 6 volts. If this range of voltage cannot be obtained, change the biasing resistor R_E until the appropriate voltage is achieved.

2. When the voltage is within the desired range, make the necessary measurements to complete the Actual Test Results row of Table 6-1. Note that currents I_E and I_C must be calculated from the dc voltages measured for Table 6-1 and the formulas given in step A. If it was necessary to change the biasing resistor, R_E, note the change on the circuit drawn in step A.1.

D. 1. Connect an oscilloscope across the output, V_{Cb}.

2. Apply an input signal of 1 kHz from a sine-wave generator and increase the input signal until the output sine wave distorts. Then decrease the input until a nondistorted sine wave appears across the output. The input signal will be less than 100 mV. Therefore, a voltage divider may be required across the signal generator.

3. Measure the input voltage, v_{eb}, and output voltage, v_{cb}, using an ac VTVM. Record the values obtained in Table 6-3, page 102. If an ac VTVM is not available, use an oscilloscope and record the peak-to-peak readings in Table 6-3.

E. Decrease the input signal to the smallest readable value (approximately 1 mV) and record the input and output voltages in Table 6-3.

TABLE 6-3 Ac Circuit Readings

Step	v_{in} in V	v_{eb} in V	v_{cb} in V	$R_L = R_O$ in Ω	A_v Calculated
D	———			———	
E	———			———	
F				———	———
G					———

F. 1. To determine the input current and input resistance of the amplifier circuit, insert a 100-Ω resistor before capacitor C_1, as shown in figure 6-7.

2. Remove R_L in the circuit of figure 6-7.

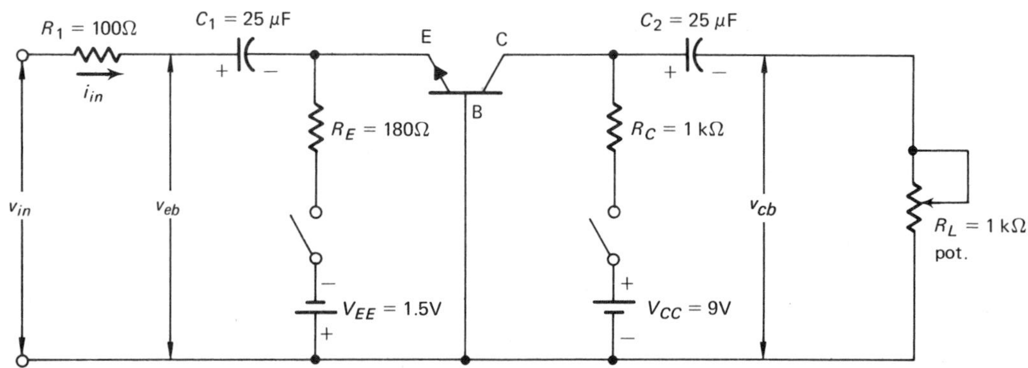

FIG. 6-7 COMMON-BASE AMPLIFIER WITH 100Ω RESISTOR TO MEASURE R_{in} AND R_L; POTENTIOMETER TO MEASURE R_O

3. Increase the input signal so that the voltage at v_{eb} has the same value as in step D.3.

4. Read and record v_{in}, v_{eb}, and v_{cb} in Table 6-3.

G. 1. To measure the output resistance of the circuit, connect the potentiometer R_L as shown in figure 6-7.

2. Adjust R_L until the output voltage v_{cb} is one-half of its value in step F.

3. Record the voltage readings in Table 6-3.

4. Remove R_L from the circuit without changing its setting. Measure R_L and record this value as R_O in Tables 6-2 and 6-3.

H. Calculate the following ac performance quantities using the readings taken for step F. (Refer to Table 6-3.) List the results in Table 6-2. If an oscilloscope is used to read the ac voltages, convert the peak-to-peak values to rms values before making any calculations.

1. Amplifier input ac current i_{in}.

$$i_{in} = \frac{v_{in} - v_{eb}}{100} \qquad i_{in} = ?$$

2. Amplifier input resistance R_{in}.

$$R_{in} = \frac{v_{eb}}{i_{in}} = ?$$

3. Ac output current i_o.

$$i_o = \frac{v_{cb}}{R_C} = ?$$

4. Voltage gain of the amplifier A_V.

$$A_V = \frac{v_{cb}}{v_{eb}} = ?$$

5. Current gain of the amplifier A_i.

$$A_i = \frac{i_o}{i_{in}} = ?$$

I. Calculate the voltage gain of the amplifier for steps D and E in Table 6-3. List the voltage gains in Table 6-3 in the column marked A_V.

DISCUSSION QUESTIONS

1. Assuming that V_{CB} was 2 V in step C, answer and discuss the following questions:
   a. Find the values of I_C and I_E.
   b. What must be done with resistor R_E to change I_C to 5 mA?
   c. Can the power supply V_{EE} be changed instead of R_E so that I_C will equal 5 mA? How?
2. Compare the theoretical and actual test results listed in Table 6-1. Discuss and give reasons for any discrepancies.
3. Compare the theoretical and actual test results listed in Table 6-2. Discuss and give reasons for any discrepancies.
4. The power gain of the amplifier is determined by multiplying the voltage gain by the current gain. What is the power gain of the CB amplifier circuit? Why is the power gain approximately equal to the voltage gain?
5. List some general conclusions about the CB amplifier. For what applications can it be used? For what applications is it unsuitable?
6. Why are the voltage gains in steps D and E of Table 6-3 different? Will this difference be true for all amplifiers?

EXTENDED STUDY TOPICS

1. Using the transistor characteristics shown in figure 6-3a, find the missing quantities in the following table.

α	I_C in mA	I_B in mA	I_E in mA	I_{CBO} in μA	V_{CB} in V
			20	0	-5
			10	0	-10
			20	4.2	-7.5
			30	8.7	-3

2. Find the missing quantities in the following table.

α	I_C in mA	I_B in mA	I_E in mA	I_{CBO} in μA
0.95		0.4		5
0.95			6.7	10
0.91			150	100
	10	1.0		8

3. A CB transistor experiences a change in V_{EB} of 1.5 mV when the emitter current changes 0.01 mA. Find the amplifier input resistance of $R_E = 500\ \Omega$.

4. Find the voltage gain of the CB transistor amplifier in Topic 3, if $\alpha = 0.95$, $I_C = 2$ mA, and the voltage across resistor R_C is 20 V.

5. A CB amplifier circuit has $I_C = 20$ mA, and the voltage across resistor R_C is 35 V. Find the amplifier output resistance if $r_{ob} = 1\ M\ \Omega$.

6. Find the approximate voltage gain of the CB amplifier in Topic 5 if $r_{ib} = 50\ \Omega$.

Unit 7

Common-emitter and common-collector amplifiers

OBJECTIVES

After studying this unit, the student will be able to discuss and demonstrate an understanding of the basic principles of:

- Common-emitter and common-collector transistor configurations.
- Common-emitter transistor characteristics.
- Calculating β (beta) from the common-emitter transistor characteristics or from data supplied from a common-emitter transistor circuit.
- Connecting a common-emitter and a common-collector transistor circuit and taking data to explain the characteristics of these types of circuits.

COMMON-EMITTER CHARACTERISTIC CURVES

The circuits in figure 7-1 illustrate the biasing of PNP and NPN transistors with common-emitter (CE) configurations. The *emitter* is the *common* terminal between the base input terminal and the collector output terminal. The parameters V_{BE}, I_B, V_{CE}, and I_C can be measured at the indicated points in the circuits of figure 7-1. The base current I_B is the input current and the collector current I_C is the output current. The input voltage is V_{BE} and the output voltage is V_{CE}. To study the behavior of the CE transistor, it is recommended that the relationship between I_B and V_{BE} be graphed; the resulting curves are called the *input characteristic.* In addition, the relationship between I_C and V_{CE} can be graphed resulting in curves called the *output characteristic.*

To determine the CE input characteristic for an NPN transistor, it is necessary to vary and record the input voltage, V_{BE}, and the input current, I_B, for different values of the collector-emitter voltage V_{CE}. For a given value of V_{CE}, the graph of I_B versus V_{BE} is essentially the same as that of a *p-n* junction diode, figure 7-2. (Of course, the *p-n* junction diode is the emitter-base junction diode.)

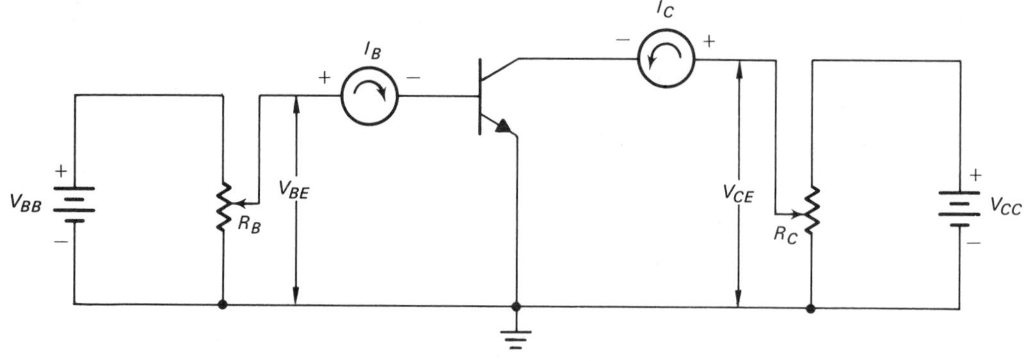

(a) NPN TRANSISTOR

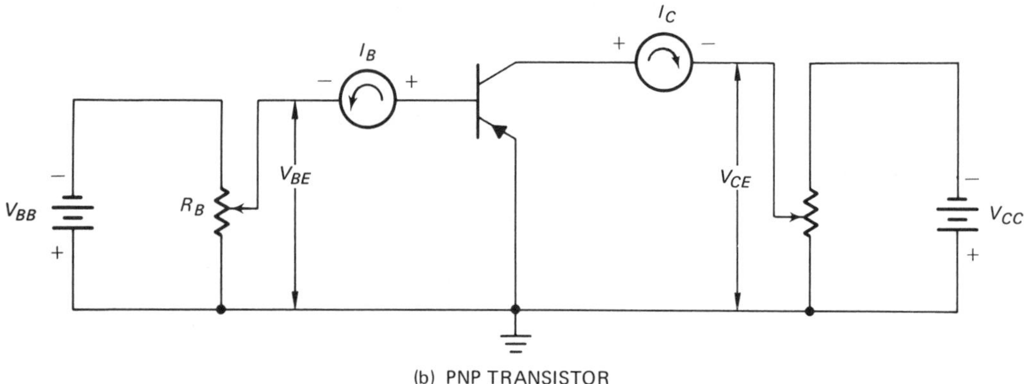

(b) PNP TRANSISTOR

FIG. 7-1 CIRCUITS USED TO OBTAIN THE CE CHARACTERISTICS

Increases in V_{CE} in figure 7-2 cause the emitter-base junction diode to become a slightly poorer diode. In other words, I_B becomes slightly larger for a given V_{BE} as V_{CE} increases. Thus, the variation of output voltage V_{CE} has little effect upon the voltage across the emitter-base junction. In fact, the input curves will show little variation for

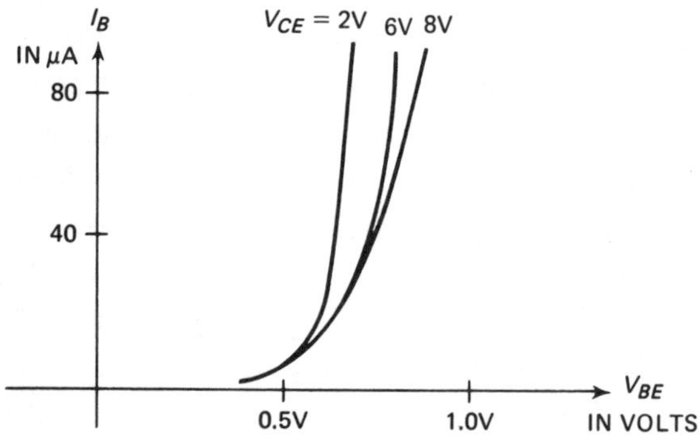

FIG. 7-2 COMMON-EMITTER INPUT CHARACTERISTICS FOR A TYPICAL NPN SILICON TRANSISTOR

different transistors. The voltage values listed in Table 5-1 for the CB voltage V_{EB} also apply to the CE voltage V_{BE}.

The characteristic curves which are most useful for a transistor in the CE configuration are the *collector* or *output* characteristics, figure 7-3. Figure 7-3a illustrates a typical set of curves for a PNP transistor, and figure 7-3b illustrates a set of curves for an NPN transistor.

To determine the CE collector characteristics for an NPN transistor, the following steps are performed using the circuit of figure 7-1a.

1. The input circuit, I_B, is adjusted to a particular value. This value is recorded and is held constant.
2. The output voltage, V_{CE}, is varied and recorded.

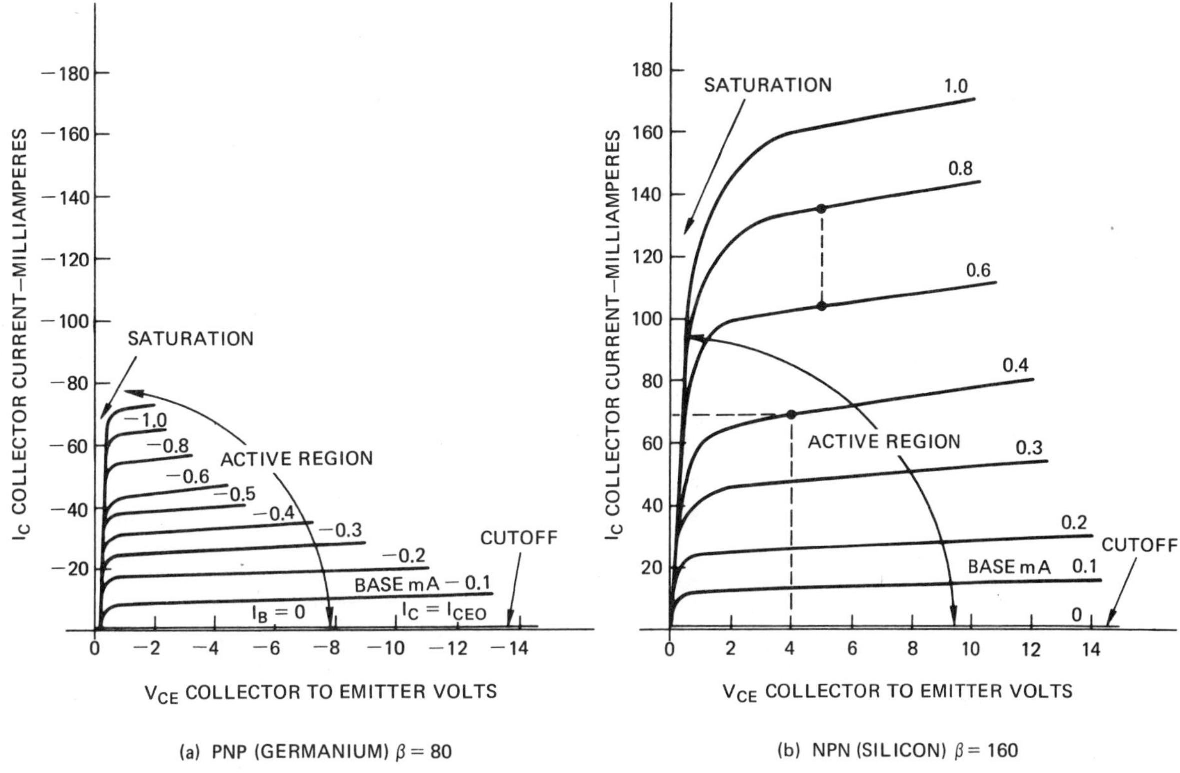

FIG. 7-3 TYPICAL CE COLLECTOR CHARACTERISTICS FOR PNP AND NPN TRANSISTORS

3. The output current, I_C, is recorded as the output voltage changes.
4. Steps 1,2, and 3 are repeated for different values of input current.

The resulting graph of I_C versus V_{CE} for various values of I_B is shown in figure 73-b. These characteristic curves can be used to determine the dc and ac current gain of a transistor in the CE configuration (using Eq. 7.5 and Eq. 7.7).

Although the physical operation of the CB and CE transistor circuits is identical, new active region current equations can be developed for the CE amplifier. According to the CE equations in the active region,

$$I_B = (1 - \alpha)\, I_E - I_{CBO} \qquad \text{Eq. 5.6}$$

Solving for I_E yields:

$$I_E = \frac{I_B + I_{CBO}}{(1 - \alpha)}$$

Substituting Eq. 5.1 for I_E ($I_E = I_C + I_B$) in the previous equation gives:

$$I_C + I_B = \frac{I_B + I_{CBO}}{(1 - \alpha)}$$

By rearranging the terms and solving for I_C, the resulting equation is:

$$I_C = \frac{\alpha}{1 - \alpha} I_B + \frac{1}{1 - \alpha} I_{CBO} \qquad \text{Eq. 7.1}$$

The CE current gain, β (beta) (or h_{FE}), is defined from Eq. 7.1 as:

$$\beta = \frac{\alpha}{1 - \alpha} \qquad \text{Eq. 7.2}$$

Also, from Eq. 7.1, the CE reverse leakage current, I_{CEO}, is defined as:

$$I_{CEO} = \frac{I_{CBO}}{1 - \alpha} \qquad \text{Eq. 7.3}$$

Substituting Eq. 7.2 and Eq. 7.3 into Eq. 7.1 and yields:

$$I_C = \beta\, I_B + I_{CEO} \qquad \text{Eq. 7.4}$$

Equation 7.4 says that the collector current is equal to the CE current gain times the base current plus a leakage current. The leakage current, I_{CEO}, is the current flowing in the collector if the base current is zero.

When dealing with ac quantities, the CE current gain is defined as:

$$\beta_{ac} = \frac{\Delta I_C}{\Delta I_B}\Bigg|_{V_{CE} \text{ is constant}} \qquad \text{Eq. 7.5}$$

If Eq. 7.4 is solved for β, then we have:

$$\beta = \frac{I_C - I_{CEO}}{I_B} \qquad \text{Eq. 7.6}$$

If the CE leakage current is small compared to the collector current, Eq. 7.6 becomes:

$$\beta = \frac{I_C}{I_B} \qquad \text{Eq. 7.7}$$

According to Eq. 7.7, β is the ratio of the collector (output) current to the base (input) current.

By substituting Eq. 7.4 into Eq. 5.1 for I_C, I_E can be determined in terms of I_B and β.

$$I_E = I_B + (I_C)$$
$$= I_B + (\beta + I_{CEO})$$
$$I_E = (\beta + 1)\, I_B + I_{CEO} \qquad \text{Eq. 7.8}$$

Eq. 7.8 shows that the emitter current is essentially $(\beta + 1)$ times the base current plus I_{CEO}.

A relationship between α and β can be derived as follows:

$$\beta = \frac{\alpha}{1 - \alpha} \qquad \text{Eq. 7.2}$$

Solving for α in terms of β yields:

$$\alpha = \frac{\beta}{\beta + 1} \qquad \text{Eq. 7.9}$$

The relation of $(\beta + 1)$ to α is also important. From Eq. 7.2:

$$\beta + 1 = \frac{\alpha}{1 - \alpha} + 1$$

$$\beta + 1 = \frac{1}{1 - \alpha} \qquad \text{Eq. 7.10}$$

The above relationships hold for both dc and ac current gains.

PROBLEM 1.

A transistor has $\beta = 100$ and negligible leakage current. Find α.

$$\alpha = \frac{\beta}{\beta + 1} = \frac{100}{101} = \mathbf{0.99}$$

PROBLEM 2.

The transistor in Problem 1 is used as a CB amplifier in the active region. The emitter current is 1mA. What is the collector current?

$$I_C = \alpha I_E + I_{CBO}$$
$$= 0.99 \times 1 \text{ mA} + 0.0 = \mathbf{0.99 \text{ mA}}$$

PROBLEM 3.

If a transistor has $\alpha = 0.95$ and negligible leakage current, what is β?

$$\beta = \frac{\alpha}{1 - \alpha} = \frac{0.95}{1 - 0.95} = \frac{0.95}{0.05} = \mathbf{19}$$

PROBLEM 4.

The transistor in Problem 3 is used as a CE transistor in the active region. The base current is 0.1 mA. What is the collector current?

$$I_C = \beta I_B + I_{CEO}$$
$$= 19 \times 0.1 \text{ mA} + 0.0 = \mathbf{1.9 \text{ mA}}$$

PROBLEM 5.

A transistor has the collector characteristics shown in figure 7-3b. Calculate the β at $V_{CE} = 4$ V and $I_B = 0.4$ mA. Find the emitter current for these conditions.

$$\beta = \frac{I_C}{I_B}\Bigg|_{V_{CE} = 4 \text{ V}}$$
$$= \frac{69.0 \text{ mA}}{0.4 \text{ mA}} = 173$$

$$I_E = (\beta + 1)\, I_B + I_{CEO}$$
$$= (173 + 1)\, 0.4 \text{ mA} + 0.0$$
$$= 174 \times 0.4 \text{ mA}$$
$$= \mathbf{69.6 \text{ mA}}$$

PROBLEM 6.

A transistor has the collector characteristics shown in figure 7-3b. Calculate the β_{ac} at V_{CE} = 5 V, with a base current change of 0.6 mA to 0.8 mA. Find the emitter current change for these conditions.

$$\beta_{ac} = \frac{\Delta I_C}{\Delta I_B} \Bigg|_{V_{CE} = 5\ V}$$

$$= \frac{(135 - 104)\ mA}{(0.8 - 0.6)\ mA} = \frac{31.0\ mA}{0.2\ mA} = \mathbf{155}$$

$$I_E = (\beta + 1)\ \Delta I_B + I_{CEO}$$

$$= (155 + 1)\ (0.8 - 0.6)\ mA + 0.0$$

$$= 156 \times 0.2\ mA = \mathbf{31.2\ mA}$$

The set of CE characteristic curves for NPN and PNP transistors shown in figure 7-3 are typical. If manufacturers provide any characteristic curves, it is usually those for the CE configuration. The CE characteristic curves show that little base current flows for a considerable amount of collector current. In other words, the CE current gain is a large number. The CE current gain will vary between 10 to 500, depending upon the particular transistor. In *super-beta* transistors, the current gain may reach a magnitude of 10^4.

A study of the curves in figure 7-3 shows that the collector current is not zero. When I_B is zero, the only current flowing in the collector circuit is the *leakage current*, I_{CEO}. This region of operation is called the *cutoff region*.

If the transistor is to operate in the cutoff region, then the emitter-base junction must have a zero bias or must be reverse biased. Thus, the only current flowing in the collector circuit is the very small leakage current, I_{CEO}, which is usually expressed in μA. The output voltage, V_{CE}, will be very close in value to the supply voltage since the voltage drop across the collector resistor is very small.

When the emitter-base junction is forward biased, then base current flows and the collector current is equal to β times the base current. This region of operation is called the *active region* and is shown in figure 7-3. The current equations presented earlier in this unit are used in the active region.

If the base current is increased further, a point is reached at which the collector current will no longer increase. This region is called the *saturation region* and is shown in figure 7-3. In the saturation region, the voltage across the collector-emitter is nearly equal to zero volts, but the collector-base junction is slightly forward biased. For example, an NPN germanium transistor has a typical saturation voltage, V_{CE}, of +0.1 V. The approximation for the collector current in the saturation region is expressed by Eq. 5.7.

Each of the transistor configurations has the same three possible regions of operation, namely, the *active, saturation,* and *cutoff* regions. Figure 7-4, page 110, shows the CE biasing required and the voltages that exist across an NPN transistor for each region of operation.

Using the transistor characteristics of figure 7-3b, find the missing quantities in the following table. (R 7-1)

β	I_C in mA	I_B in mA	I_E in mA	I_{CEO} in μA	V_{CE} in V
		0.4		0	2
		1.0		0	5
		0.8		10	8
		0.4		50	9

SIMPLE COMMON-EMITTER AMPLIFIERS

Typical CE amplifier circuits biased in the active region are shown in figure 7-5, page 111. The input dc supply is V_{BB} and the collector supply is V_{CC}.

Figure 7-6, page 111, illustrates another method of biasing a CE amplifier using only one power supply. For this circuit the base current is varied by adjusting the resistor R_B.

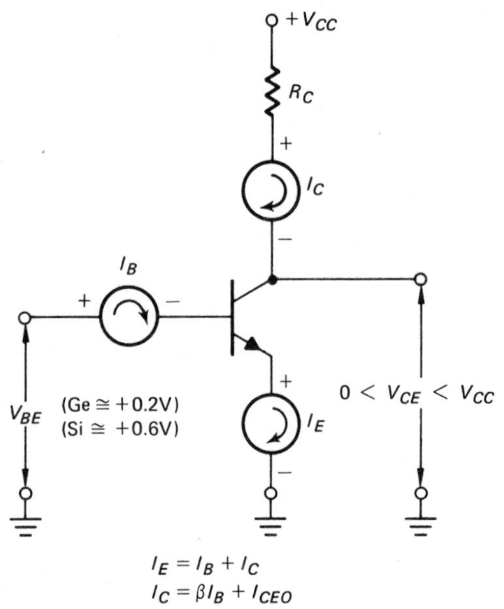

$I_E = I_B + I_C$
$I_C = \beta I_B + I_{CEO}$

ACTIVE REGION OPERATION WITH EMITTER-BASE JUNCTION *FORWARD* BIASED AND COLLECTOR-BASE JUNCTION *REVERSE* BIASED.

(a)

+V_{CC}
R_C
I_C
I_B
V_{BE}
I_E
$V_{CE} \cong 0$

$I_C \cong I_{CS} \cong \frac{V_{CC}}{R_C}$

SATURATION REGION OPERATION WITH EMITTER-BASE JUNCTION *FORWARD* BIASED AND COLLECTOR-EMITTER VOLTAGE NEARLY EQUAL TO ZERO VOLTS.

(b)

+V_{CC}
R_C
I_C
I_B
$V_{BE} = 0$ OR SLIGHTLY NEGATIVE
I_E
$V_{CE} \cong V_{CC}$

$I_B = 0$
$I_C = I_{CEO} = (\beta + 1)\, I_{CBO}$ (LEAKAGE CURRENT)

CUTOFF REGION OPERATION WITH EMITTER-BASE JUNCTION *REVERSE* BIASED AND COLLECTOR-BASE JUNCTION *REVERSE* BIASED.

(c)

FIG. 7-4 THREE REGIONS OF OPERATION FOR A COMMON-EMITTER NPN TRANSISTOR

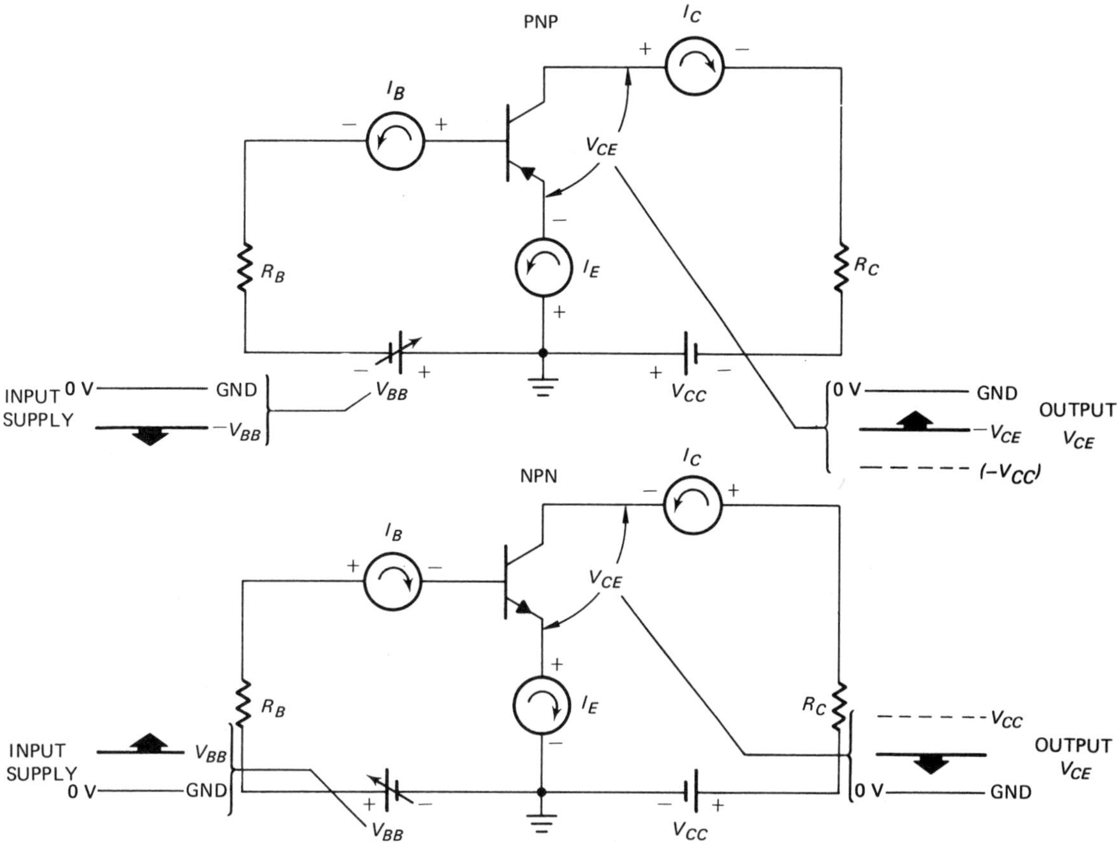

FIG. 7-5 COMMON-EMITTER AMPLIFIER CIRCUIT SHOWING THE EFFECT OF CHANGES OF THE INPUT SUPPLY ON THE OUTPUT VOLTAGE

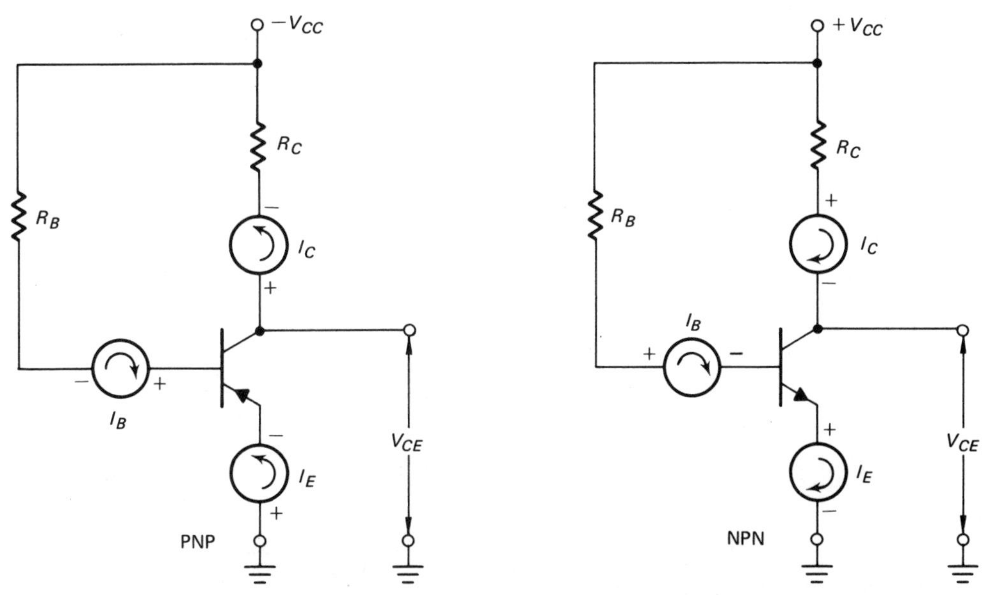

FIG. 7-6 COMMON-EMITTER CONFIGURATION FOR THE ACTIVE REGION (ONE POWER SUPPLY IS USED TO BIAS BOTH THE COLLECTOR AND THE BASE CIRCUITS)

If V_{BB} is increased in the circuits shown in figure 7-5, then the base current is increased. Since there is little change in the base-emitter voltage drop, the increase in V_{BB} is dropped across the base resistor R_B. As long as the transistor operation remains in the active region, the collector current will increase due to the fact that $I_C = \beta I_B + I_{CEO}$. When the collector current increases, the collector-emitter voltage drop decreases since the voltage drop across the collector resistor R_C has increased. For the PNP transistor shown in figure 7-5, as the base becomes more negative, V_{CE} will become less negative (or more positive). For the NPN transistor shown in figure 7-5, as the base becomes more positive, V_{CE} will become less positive. If the supply voltage V_{BB} is decreased, then the opposite of the above effects will occur.

If V_{BB} remains constant and V_{CC} is increased, with the transistor operation remaining in the active region, then the value of the collector-emitter voltage, V_{CE}, increases proportionally. The collector current, however, shows hardly any change. The common-emitter characteristics, in figure 7-3 show that the collector current increases only slightly after the initial reverse bias is applied. If the input supply V_{BB} is increased considerably, the transistor will move into saturation, as in the common-base transistor circuit. In this region, the collector current is approximately V_{CC}/R_C and the collector-emitter voltage is approximately zero. If the polarity of V_{BB} is reversed so that the emitter-base junction is reverse biased, the transistor will be cut off. For this condition, the output voltage will approximately equal the supply voltage V_{CC}.

PROBLEM 7.

For the circuit shown in figure 7-7, determine the following: the maximum collector current that can flow, the voltage from the collector terminal to ground if point A opens, and the voltage from the collector terminal to ground if point B opens.

The maximum collector current occurs when the total supply voltage (9V) is dropped across the 1-kΩ collector resistor. For this circuit, the voltage from the collector to the emitter must be zero volts. As a result, the transistor is saturated.

$$I_{CS} = \frac{9\text{ V}}{1\text{ k}\Omega} = 9.0\text{ mA}$$

With point A open, a complete circuit does not exist between the 9-V supply and the collector terminal. Therefore, no current flows through the 1-kΩ resistor, resulting in zero voltage across the 1-kΩ resistor. In other words, the voltage from the collector to the ground is equal to the 9-V supply voltage. With point B open, the base-emitter circuit is not complete. Thus, the base current is equal to zero. However, the collector-emitter junction is reverse biased because of the 9-V supply and the current flowing in the collector-emitter circuit is the leakage current I_{CEO}. I_{CEO} is small. As a result, the voltage drop across the 1-kΩ resistor is small and the voltage from the collector terminal to ground is approximately equal to the 9-V supply voltage.

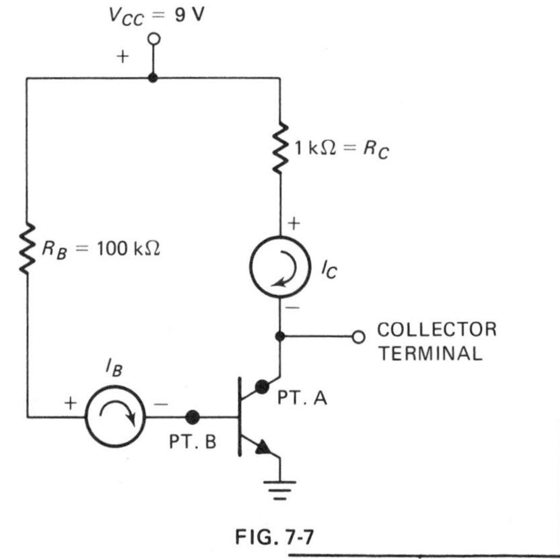

FIG. 7-7

PROBLEM 8

For the circuit shown in figure 7-8, the emitter current is equal to 1 mA and a silicon transistor is used. Determine the following voltages: the emitter to ground voltage, V_E; the collector to ground voltage, V_C; the collector to emitter voltage, V_{CE}; and the base to ground voltage, V_B.

V_E: The emitter current is equal to 1mA. Using Ohm's Law,

$$V_E = I_E R_E = 1 \text{ mA} \times 1 \text{ k}\Omega = \mathbf{1\ V}$$

V_C: Since neither current gain is given, assume that $\alpha = 1$; thus, $I_C = I_E$. Using Kirchhoff's voltage law.

$$V_C = V_{CC} - I_C R_C = 15 \text{ V} - 1 \text{ mA} \times 8 \text{ k}\Omega$$
$$= 15 \text{ V} = 8 \text{ V} = 7 \text{ V}$$

V_{CE}: Using Kirchhoff's voltage law,

$$V_{CE} = V_C - V_E = 7 \text{ V} - 1 \text{ V} = \mathbf{6\ V}$$

V_B: Due to the fact that the transistor is silicon, the voltage across the base-emitter junction, V_{BE}, is equal to 0.6 V. The voltage from the base to ground is found using Kirchhoff's voltage law.

$$V_B = V_E + V_{BE} = 1.0 \text{ V} + 0.6 \text{ V} = \mathbf{1.6\ V}$$

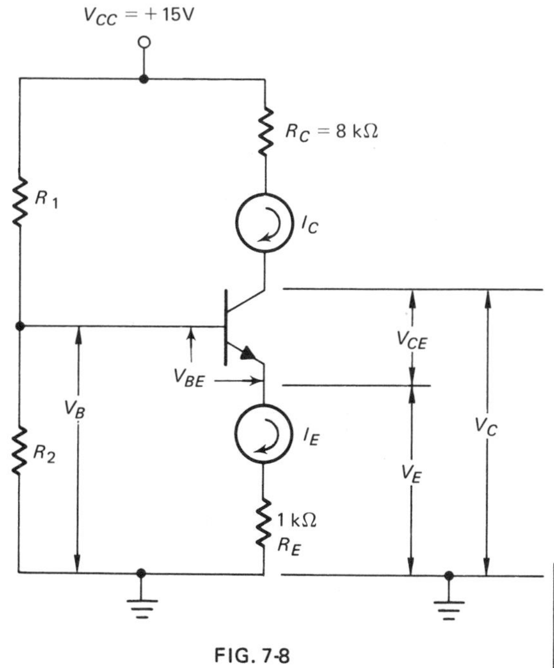

FIG. 7-8

This unit has presented the input and output characteristic curves for PNP and NPN transistors in the CE configuration. Certain parameters can be obtained from these curves. One parameter is the current gain. Two other parameters for CE transistors that can be obtained from these curves are the input resistance, r_{ie}, and the output resistance, r_{oe}. These parameters are expressed mathematically as follows:

$$r_{ie} = \frac{\Delta V_{BE}}{\Delta I_B} \qquad \text{Eq. 7.11}$$

where ΔV_{BE} = change in base-emitter voltage

ΔI_B = change in base current

r_{ie} = input resistance of a CE transistor

$$r_{oe} = \frac{\Delta V_{CE}}{\Delta I_C} \qquad \text{Eq. 7.12}$$

where ΔV_{CE} = change in collector-emitter voltage

ΔI_C = change in collector current

r_{oe} = output resistance of a CE transistor

The following problems illustrate the approximate values for r_{ie} and r_{oe} for CE transistors.

PROBLEM 9

A transistor experiences a change in V_{BE} of 2 mV when the base current change is 1 μ A. Find the resistance r_{ie}.

$$r_{ie} = \frac{\Delta V_{BE}}{\Delta I_B} = \frac{2.0 \text{ mV}}{1.0\,\mu\text{A}} = \mathbf{2000\ ohms}$$

Problem 9 shows the CE transistor input resistance which, typically, ranges between 1000 and 5000 ohms. Since the emitter current is $(\beta + 1)$ times larger than the base current, the resistance looking into the base should be $(\beta + 1)$ larger than the resistance looking into the emitter. Mathematically, this statement is expressed as:

$$r_{ie} = (\beta + 1)\, r_{ib} \qquad \text{Eq. 7.13}$$

PROBLEM 10

A transistor experiences a change in V_{CE} of 15 V when the collector current changes by 10μ A. Find the resistance r_{oe}.

$$r_{oe} = \frac{\Delta V_{CE}}{\Delta I_C} = \frac{15\ \text{V}}{10\ \mu\text{A}} = 1.5\ \text{M}\,\Omega$$

Problem 10 shows that the output resistance for the CE transistor is large due to the fact that the collector-emitter junction is reverse biased. Typically, the output resistance is greater than 1 M Ω, although it is not quite as large as the value for the CB configuration.

In figure 7-9 the CE transistor resistance r_{ie} is in parallel with the base resistance R_B. This parallel combination of resistances is the *amplifier input resistance.* This is the resistance seen by a voltage source if it is connected to the input of the circuit in figure 7-9.

PROBLEM 11

A CE amplifier has an R_B value of 8000 ohms. The amplifier transistor has an r_{ie} of 2000 ohms. Find the amplifier input resistance.

$$R_{in} = r_{ie} \,||\, R_B = 2000\ \Omega \,||\, 8000\ \Omega$$

$$= \frac{2000 \times 8000}{2000 + 8000} = 1600\ \Omega$$

Referring to figure 7-9 again, the transistor output resistance, r_{oe}, is in parallel with the collector resistance R_C. This parallel combination of resistances is the *amplifier output resistance.* This is the resistance seen by another transistor amplifier connected to the output of the circuit in figure 7-9. Since the value of R_C is normally much smaller than r_{oe}, the amplifier output resistance is approximately equal to R_C.

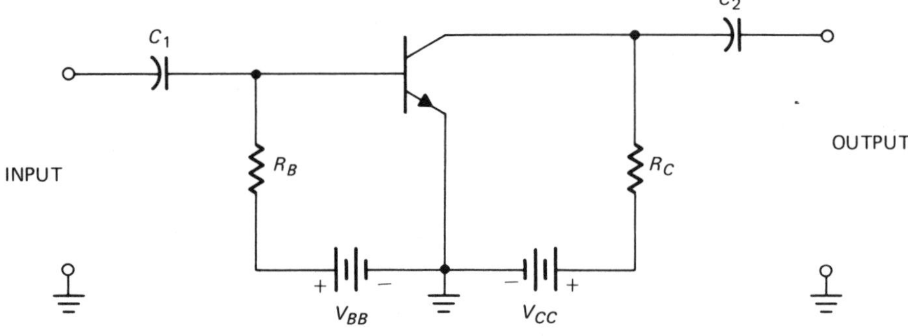

(a) COMPLETE AMPLIFIER CIRCUIT

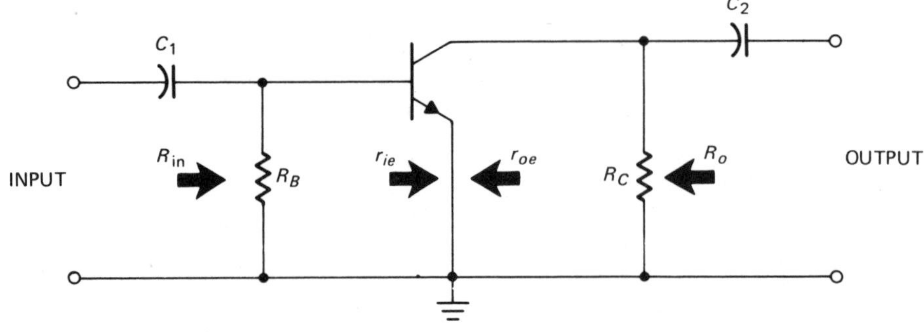

(b) AC EQUIVALENT AMPLIFIER CIRCUIT

FIG. 7-9 TYPICAL CE AMPLIFIER CIRCUIT USING AN NPN TRANSISTOR

PROBLEM 12

A CE amplifier circuit has R_C = 10,000 ohms. The amplifier transistor has an r_{oe} of 1 M Ω. Find the amplifier output resistance.

$$R_o = r_{oe} \| R_C = 1\ M\Omega \| 10000\ \Omega$$
$$= \mathbf{10000\ \Omega}$$

The voltage gain, A_V, for a CE amplifier is defined as:

$$A_V = \frac{\Delta V_{out}}{\Delta V_{in}}$$

$$= \frac{\Delta I_C\ R_o}{\Delta I_{in}\ R_{in}}$$

For the CE amplifier, the following approximations can be made in the previous equation.

$$\Delta I_C \cong \beta \Delta I_B$$

$$I_B = \frac{R_B}{R_B + r_{ie}} \times I_{in}$$

$$R_o \cong R_C$$

$$R_{in} \cong r_{ie} \| R_B$$

Substituting the above approximations into the equation for A_V yields:

$$A_V \cong \frac{\beta \Delta I_B\ R_C}{\Delta I_{in}\ (r_{ie} \| R_B)}$$

$$= \frac{\beta \Delta I_B\ R_C}{\left[\frac{(R_B + r_{ie})\Delta I_B}{R_B}\right]\left[\frac{r_{ie} \times R_B}{r_{ie} + R_B}\right]}$$

$$A_{V\,(CE)} \cong \frac{\beta R_c}{r_{ie}} = \frac{\beta\ R_C}{(\beta + 1)\ r_{ib}}$$

and

$$A_{V\,(CE)} \cong \frac{R_C}{r_{ib}} \qquad \textbf{Eq. 7.14}$$

Eq. 7.14 is only an *approximation* of the CE voltage gain. However,this approximation is sufficient for this analysis.

PROBLEM 13

For a circuit, the transistor and parameters are the same as those of Problems 11 and 12. The CE current gain is 32. Find the approximate voltage gain.

$$A_V = \frac{\beta R_C}{r_{ie}} + \frac{32 \times 10000}{2000} = \mathbf{160}$$

Problem 13 shows that the CE amplifier configuration has a current gain and a voltage gain which are both greater than unity. Thus, the CE amplifier is an excellent power amplifier and most amplifier circuits use the CE configuration.

In the circuit in figure 7-6, R_C = 2.2 kΩ and V_{CC} = 10 V. Find the collector saturation current. If β = 100, what is the base current if I_{CBO} = 0? (R7-2)

SIMPLE COMMON-COLLECTOR AMPLIFIERS

The common-collector (CC) configuration is similar to the CE configuration with the exception that the output is taken at the emitter rather than at the collector. The characteristic curves for the CC configuration are identical to the CE characteristics. In addition, the CE current equations are used for the CC circuit. A typical CC circuit is shown in figure 7-10, page 116.

The CC output voltage is always equal to the input voltage minus the base-emitter voltage drop. Since the output voltage follows the input voltage closely, the CC configuration is called an *emitter follower.*

The emitter follower transistor configuration is commonly used as an impedance matching device. Because of its high input resistance and low output resistance, it can replace transformers in certain applications. In addition to its impedance matching properties, it has excellent frequency response. Although the emitter follower has no *voltage* gain, it can be used as a current amplifier because of its high current gain. The emitter follower is widely used as a buffer element or current amplifier in integrated circuits.

A buffer element is used to isolate one part of a circuit from another part. Since most integrated circuits are directly coupled between stages, these devices use emitter followers to prevent one stage from loading down the previous stage.

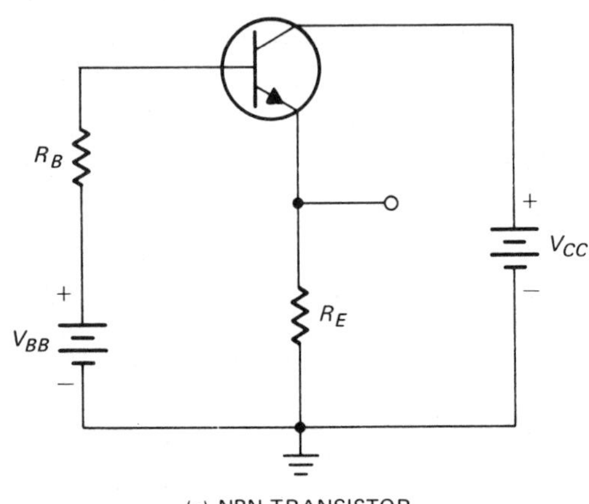

(a) NPN TRANSISTOR

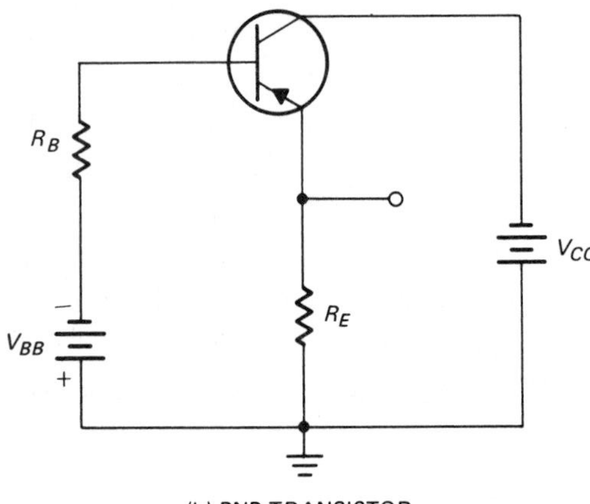

(b) PNP TRANSISTOR

FIG. 7-10 TYPICAL CC TRANSISTOR CONFIGURATION

PROBLEM 14

For the circuit shown in figure 7-11, the emitter current is equal to 2 mA. A silicon transistor is used. Determine the collector to emitter voltage, V_{CE}; the base to ground voltage, V_B; the base current, I_B; and the collector current, I_C.

V_{CE}: The collector to emitter voltage is found by applying Kirchhoff's voltage law.

$$V_{EE} + V_{CC} = V_{CE} + I_E R_E$$

$$V_{CE} = V_{EE} + V_{CC} - I_E R_E$$

$$= 15\text{ V} + 15\text{ V} - (2\text{ mA} \times 5\text{ k}\Omega)$$

$$= 30\text{ V} - 10\text{ V}$$

$$= \mathbf{20\text{ V}}$$

V_B: For a silicon transistor, $V_{BE} = 0.6$ V. V_B is found by applying Kirchhoff's voltage law.

$$V_{EE} = V_B + V_{BE} + I_E R_E$$

$$V_B = V_{EE} - V_{BE} - I_E R_E$$

$$= 15\text{ V} - 0.6\text{ V} - 10\text{ V}$$

$$= \mathbf{4.4\text{ V}}$$

I_B: Use Ohm's Law to find the base current.

$$I_B = \frac{V_B}{R_B} = \frac{4.4\text{ V}}{100\text{ k}\Omega} = 0.044\text{ mA}$$

$$= \mathbf{44\ \mu A}$$

I_C: The collector current is found using the transistor equation.

$$I_C = I_E - I_B = 2.0\text{ mA} - 0.044\text{ mA}$$

$$= \mathbf{1.956\text{ mA}}$$

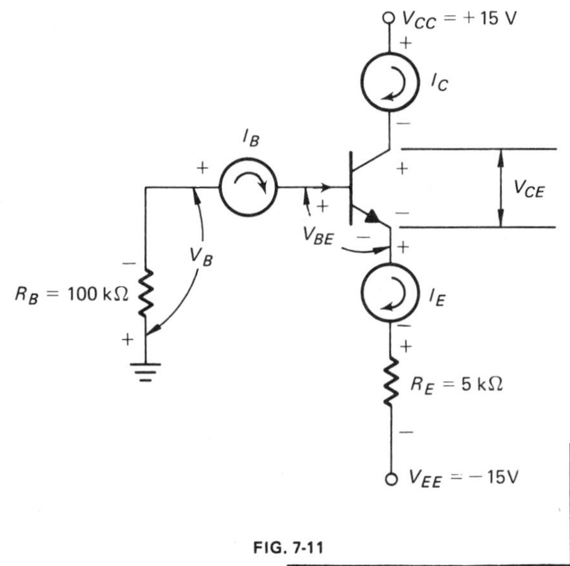

FIG. 7-11

The circuit in figure 7-12 is used to illustrate the procedure for obtaining the CC input resistance r_{ic}. First, r_{ic} is defined as follows:

$$r_{ic} = \frac{\Delta V_B}{\Delta I_B}$$

The voltage V_B is found by applying Kirchhoff's voltage law from the base to ground.

$$\Delta V_B = \Delta V_{BE} + \Delta V_E$$

Substituting the above equation into the equation for r_{ic} yields:

$$r_{ic} = \frac{\Delta V_{BE}}{\Delta I_B} + \frac{\Delta V_E}{\Delta I_B}$$

$$= \frac{\Delta V_{BE}}{\Delta I_B} + \frac{\Delta I_E R_E}{\Delta I_B}$$

The first term in the equation for r_{ic} is equal to r_{ie}. In the second term of the equation, the current ratio I_E/I_B can be replaced by $\beta + 1$ (or, as an approximation, β). The CC input resistance equation becomes:

$$r_{ic} = r_{ie} + \beta R_E \qquad \text{Eq. 7.15}$$

PROBLEM 15.

An emitter follower has the following parameters, $\beta = 100$, $R_E = 10{,}000$ ohms, and $r_{ie} = 1000$ ohms. Find the resistance r_{ic}.

$$r_{ic} = r_{ie} + \beta R_E$$

$$R_{ic} = 1000\ \Omega + 100 \times 10{,}000\ \Omega$$

$$= 1{,}000{,}000\ \Omega$$

$$= 1\ \mathrm{M}\Omega$$

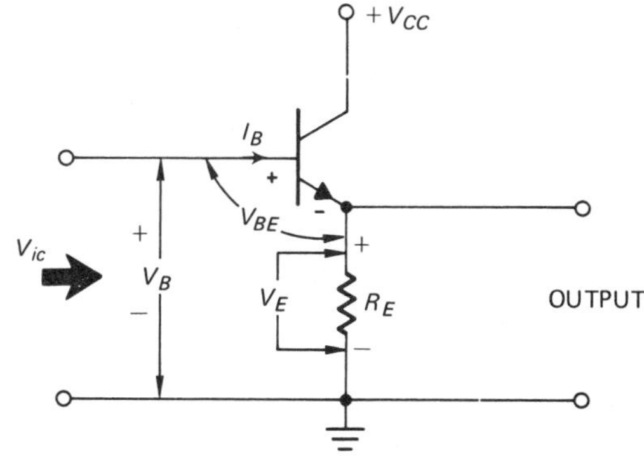

FIG. 7-12 COMMON COLLECTOR VOLTAGES

Problem 15 illustrates that the emitter follower input resistance is large. This resistance depends primarily on the CE current gain times the emitter resistance R_E. As a result, r_{ic} has a wide range of values; most of these values are very large.

The emitter follower transistor *output resistance* r_{oc} can be found by using the same equations developed for the CE output resistance r_{oe}.

In figure 7-12, the output of the emitter follower is taken across R_E. Thus, the output resistance of the emitter follower amplifier is R_E. The value of R_E is much smaller than that of the emitter follower amplifier input resistance. Because of this fact, the emitter follower makes an excellent resistance matching device.

The current gain of the emitter follower is equal to the CE current gain and varies between 10 and 500. For super-beta transistors, the magnitude of the current gain is 10^4.

The voltage gain, A_V, for the emitter follower amplifier is defined as:

$$A_V = \frac{\Delta V_{out}}{\Delta V_{in}}$$

$$= \frac{\Delta I_E R_E}{I_{in} R_{in}}$$

The following approximations can be made for the voltage gain equation for the emitter follower amplifier.

$$\Delta I_E = (\beta + 1)\ \Delta I_B$$

$$\Delta I_B = \frac{R_B}{r_{ic} + R_B}\ \Delta I_{in}$$

$$R_{in} = \frac{r_{ic} R_B}{r_{ic} + R_B}$$

Substituting these approximations into the equation for A_V yields:

$$A_V = \frac{(\beta + 1)\ \Delta I_B R_E}{\left[\frac{r_{ic} + R_B}{R_B}\right] \Delta I_B \left[\frac{r_{ic} R_B}{r_{ic} + R_B}\right]}$$

$$A_{V(cc)} = \frac{(\beta + 1) R_E}{r_{ic}} \qquad \text{Eq. 7.16}$$

In Eq. 7.16, the numerator and denominator are approximately equal; therefore, the emitter follower voltage gain has a value of slightly less than unity.

$$A_{V(cc)} < 1$$

The output voltage of a common collector circuit always equals the input voltage minus the base-emitter voltage drop.

Change R_E to 4 k Ω in Problem 14. Repeat Problem 14. (R7-3)

LABORATORY EXERCISE 7-1: CE COLLECTOR CHARACTERISTICS

PURPOSE

- To obtain the CE collector characteristics for an NPN transistor.
- To plot the CE collector characteristics.
- To determine the current gain of the CE configuration using the plot of the CE collector characteristics

MATERIALS

1 NPN transistor, 2N4425, 2N3405 or equivalent
1 Battery, 1.5 V, or dc power supply
1 Dc power supply, 12 V
2 Multirange milliammeters (or 20,000 ohm per volt VOMs)
1 VTVM, solid-state voltmeter
1 Linear potentiometer, 1 k Ω, 2 W
1 Linear potentiometer, 5 k Ω, 2 W
2 Switches, SPST

PROCEDURE

A. 1. Find the correct circuit in figure 7-1 to use in obtaining the CE characteristics for an NPN transistor. Redraw the circuit and insert a 12-V power supply for V_{CC}, a 1.5-V power supply for V_{BB}, a 1-kΩ potentiometer for R_C, a 5-kΩ potentiometer for R_B, and a switch by each power supply.

2. Find the maximum collector power dissipation rating for the transistor, P_C, in watts, For this experiment, limit the actual P_C to two-thirds the maximum P_C. Using the experimental limit P_C, calculate the collector current for each value of the collector-emitter voltage shown in Table 7-1.

TABLE 7-1

V_{CE} (volts)	0.5	1	2	3	4	5	6	8	10	12
$I_C\ (mA) = \frac{P_C}{V_{CE}}$										

B. 1. Connect the circuit drawn in step A.

2. Set the potentiometers so that no voltage is applied across the base-emitter and collector-emitter circuits. The switches should be open until readings are to be taken.

C. 1. To check that the circuit is operating correctly, close the switches and set the potentiometer R_B so that a base current of 50 to 100 μA is flowing.

2. Increase the collector-emitter voltage by adjusting potentiometer R_C. The collector current should increase to β times the base current (around 10 mA) and then remain constant as V_{CE} is increased further. If the circuit performs in this manner, then the CE characteristics can be run. If the circuit performance is not as described, consult the instructor.

D. 1. Set potentiometer R_B so that V_{BE} equals 0 volts. At this time, I_B is equal to 0 mA (I_B may be set equal to zero by removing both V_{BB} and R_B and shorting the base and emitter leads of the transistor together.

2. Vary V_{CE} through the range of values shown in Table 7-2 and record the collector current at each value of V_{CE}. (A microampere scale is required for the I_C reading.)

TABLE 7-2 CE COLLECTOR CHARACTERISTICS

	I_C(mA) measured at the following values of V_{CE}										
I_B in mA	0 V	0.5 V	1.0 V	2 V	3 V	4 V	5 V	6 V	7 V	9 V	11 V
0											
0.1											X
0.2										X	X
0.3									X	X	X
0.4								X	X	X	X
0.5							X	X	X	X	X
0.6						X	X	X	X	X	X
0.7					X	X	X	X	X	X	X
0.8					X	X	X	X	X	X	X
0.9				X	X	X	X	X	X	X	X
1.0				X	X	X	X	X	X	X	X

E. 1. Adjust the potentiometer R_C so that V_{CE} is equal to 0 V.

2. Adjust potentiometer R_B for the next value of I_B shown in Table 7-2.

3. Vary V_{CE} through the range shown in Table 7-2 and record the collector current for each value of V_{CE}. The base current must remain constant as V_{CE} is varied through its range; it may be necessary to readjust the potentiometer R_B as potentiometer R_C is adjusted to obtain V_{CE}.

F. Repeat step E for each value of the base current shown in Table 7-2. Do not exceed the collector currents calculated at each value of V_{CE} in step A2. If the experimental limit P_C is in the range of 100 mW, divide each of the required readings for I_B in Table 7-2 by 2. The range of I_B then is from 0, 0.05, 0.10, 0.15 to 0.5 mA.

G. 1. Using the data in Table 7-2, plot the CE collector characteristic curves on graph paper.

2. Plot the collector power dissipation curve calculated in step A.2, Table 7-1.

H. 1. If a transistor curve tracer is available, connect the transistor in the CE configuration.

2. Compare these tracer curves with those plotted in step G.

DISCUSSION QUESTIONS

1. As the collector-emitter voltage is varied through its range for a fixed value of base current, how does the collector current vary? Which has a greater effect on the collector current, the collector-base voltage or the base current?

2. Using the curves drawn in step G, calculate β_{dc} at each of the following locations:

   $V_{CE} = 5\text{ V}, I_B = 0.2\text{ mA}$

   $V_{CE} = 2\text{ V}, I_B = 0.5\text{ mA}$

   Using the calculated values of the current gain, find the emitter current for each value of I_B (neglect I_{CEO}).

3. Using the curves drawn in step G, determine the ac current gain of the transistor at a collector-emitter voltage of 4 V and an I_B change from 0.2 mA to 0.3 mA.

4. What quantity is measured in step D of the Procedure?

5. What is the current gain of a transistor in the CE configuration?

LABORATORY EXERCISE 7-2: COMMON-EMITTER AMPLIFIER

PURPOSE

- To predict the dc and ac performance of a common-emitter amplifier.
- To construct and then observe the operation of a common-emitter amplifier.
- To measure the dc operating point of a common-emitter amplifier.
- To measure the ac performance quantities of a common-emitter amplifier, such as input and output resistance and voltage and current gain.

MATERIALS

1 NPN transistor, 2N4425, 2N3405 or equivalent
1 Dc power supply, 15 V
1 VTVM, or solid-state voltmeter
1 Ac VTVM, use an oscilloscope if an ac VTVM is not available
1 Signal generator
1 Oscilloscope
2 Electrolytic capacitors, 25 μ F, 50 V
1 Electrolytic capacitor, 50 μ F, 50 V
3 Resistors, 1 k Ω, ½ W
1 Resistor, 5600 Ω, ½W
1 Resistor, 3300 Ω, ½ W
1 Linear potentiometer, 1000 Ω, 2 W
1 Switch, SPST, to connect power supply to circuit

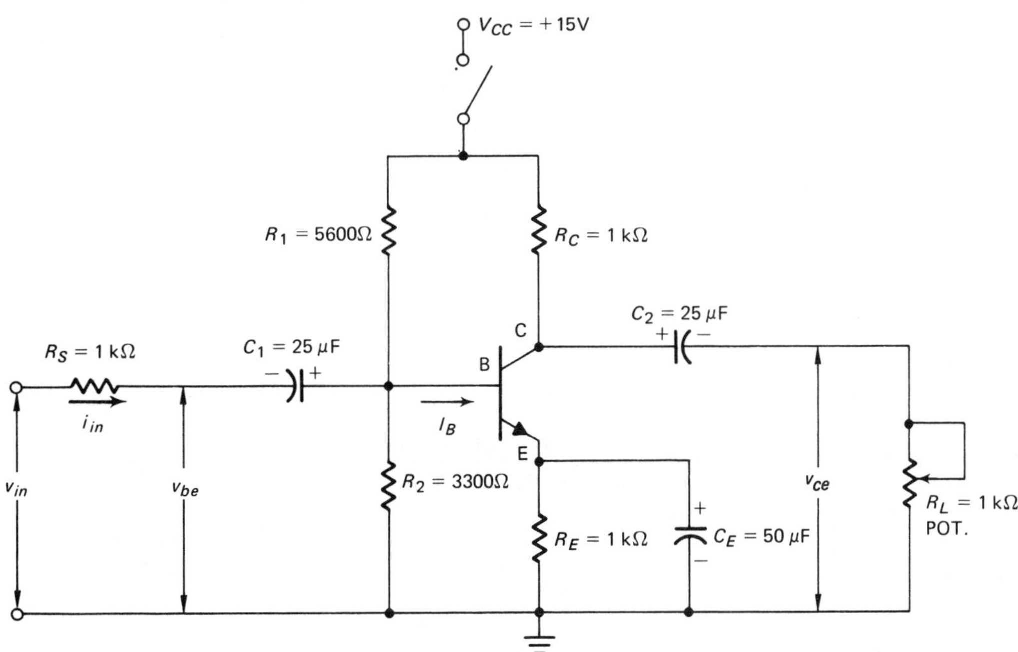

FIG. 7-13 EXPERIMENTAL COMMON-EMITTER AMPLIFIER CIRCUIT

PROCEDURE

A. 1 Convert figure 7-13, page 121, the experimental common-emitter circuit, to figure 7-9, the typical circuit. Redraw the circuit. The following relationships are explained in unit 11, CE biasing.

$$V_{BB} = \left[\frac{R_2}{R_1 + R_2}\right] V_{CC}$$

$$V_{BB} = \left[\frac{3300}{5600 + 3300}\right] \times 15\text{ V} = ?$$

$$R_B = R_1 \parallel R_2 = \frac{5600 \times 3300}{5600 + 3300} = ?$$

Resistor R_E is effectively shorted by the 50-μF capacitor in the ac equivalent circuit. However, R_E is not shorted in the dc biasing circuit. The dc voltage, (V_{EG}) is read from emitter to ground.

2. Perform the following calculations for each value of β given in Tables 7-3 and 7-4. Record the calculated values in Table 7-3. Assume that the NPN transistor is silicon, and calculate the dc base current for the experimental CE amplifier. For a silicon transistor, $V_{BE} = 0.6$ V.

$$V_{BB} = I_B R_B + V_{BE} + I_E R_E$$
$$= I_B R_B + V_{BE} + (\beta + 1) I_B R_E$$

$$I_B = \frac{V_{BB} - V_{BE}}{R_B + (\beta + 1) R_E}$$

$\beta 1 = 300,\ I_B = ?$

$\beta = 200,\ I_B = ?$

$\beta = 100,\ I_B = ?$

3. Calculate I_C and I_E for each value of I_B in step A.2. Record the values in Table 7-3.

$$I_C = \beta I_B$$

$$I_E = I_C + I_B = (\beta + 1) I_B$$

TABLE 7-3 Dc (biasing) Conditions

	β	I_B in μA	I_C in mA	I_E in mA	V_{BG} in V	V_{CG} in V	V_{EG} in V	$V_{CE} = V_{CG} - V_{EG}$ in V
Theoretical Results	300					–		
	200					–		
	100					–		
Actual Test Results	–	–						

4. Calculate voltages V_{CG}, V_{EG}, and V_{CE} for the currents calculated in step A.3. Record the values in Table 7-3.

$$V_{CG} = V_{CC} - I_C R_C$$
$$V_{EG} = I_E R_E$$
$$V_{CE} = V_{CG} - V_{EG}$$

5. Determine the input resistance, r_{ie}, of the experimental amplifier circuit, using the following equations. Record the values in Table 7-4.

$$r_{ib} \cong \frac{50\text{ mV}}{I_E}$$

$$r_{ie} \cong (\beta + 1)\, r_{ib}$$

6. Find the amplifier input resistance R_{in}. Record the values in Table 7-4.

$$R_{in} = r_{ie} || R_B$$

7. Assuming that the output resistance of the transistor, r_{oe}, is large, what is the approximate amplifier output resistance, R_o? Record the values in Table 7-4.

8. Find the approximate voltage gain for the amplifier, using the following equation. Record the values in Table 7-4.

$$A_v\,(CE) \cong \frac{R_C}{r_{ib}}$$

TABLE 7-4 Ac Performance Quantities

	β	r_{ie} in Ω	i_{in} in mA	R_{in} in Ω	i_o in mA	R_o in Ω	A_v (CE)	A_i (CE)
Theoretical Results	300		–		–			
	200		–		–			
	100		–		–			
Actual Test Results	–	–						

9. Find the approximate current gain for this amplifier. The current gain of the transistor is approximately equal to β_{ac}. However, the voltage divider drains off some of the input current. The base current to the transistor may be found by applying the current divider principle to figure 7-9b, as shown below.

$$i_b = i_{in} \frac{R_B}{R_B + r_{ie}}$$

$$R_B = R_1 \parallel R_2$$

$$A_i\,(CE) = \left[\frac{R_B}{R_B + r_{ie}}\right] \beta$$

B. 1. Connect the circuit shown in figure 7-13, but omit resistor R_S and potentiometer R_L. These resistors will be inserted in later steps. With R_S removed,

$$v_{in} = v_{be}$$

2. Check the operating point of the circuit; that is, measure the dc voltages V_{CG} and V_{EG}. The value of V_{CG} should be between 9 and 12 volts. If this voltage range cannot be obtained, change the biasing resistors in the voltage divider circuit. Try changing R_2 first. When the voltage is within the desired range, complete the Actual Test Results row of Table 7-3. Currents I_C and I_E are calculated using the dc voltages measured here and the formulas given in step A. If the biasing resistors R_1 or R_2 were changed, indicate the correct values in the circuit in figure 7-13.

C. 1. Connect an oscilloscope across the output, v_{ce}.

2. Apply a 1-kHz input signal from a sine-wave generator.

3. Increase the input signal until the output sine wave distorts. Then decrease the input until a nondistorted sine wave appears across the output. Since input signal will be less than 100 mV, a voltage divider may be required across the signal generator.

4. Measure the input voltage, v_{be}, and the output voltage, v_{ce}, with an ac VTVM. Record the values in Table 7-5. If an ac VTVM is not available, use an oscilloscope and record the peak-to-peak readings in Table 7-5.

TABLE 7-5 Ac Circuit Readings

Step	V_{in} in V	V_{be} in V	V_{ce} in V	$R_L = R_o$ in Ω	A_v Calculated
C	–			–	
D	–			–	
E				–	–
F					–

D. Decrease the input signal to the smallest readable value, approximately 1 mV, and record the input and output voltages in Table 7-5.

E. 1. To determine the input current and the input resistance of the amplifier circuit, insert resistor R_S = 1000 Ω before the capacitor as shown in figure 7-13.

2. Increase the input signal so that the voltage at v_{be} is the same as that recorded in step C4.

3. Record the values of v_{in}, v_{be}, and v_{ce} in Table 7-5.

F. 1. To measure the output resistance of the amplifier circuit, connect the potentiometer R_L as shown in figure 7-13.

2. Adjust R_L until the output voltage v_{ce} is one-half of its value in step E. Record the voltage readings in Table 7-5.

3. Remove R_L from the circuit without changing the setting. Measure R_L and record this value as R_o in Tables 7-4 and 7-5.

G. Calculate the following ac performance quantities using the values recorded in step E of Table 7-5. List the results in Table 7-4. If an oscilloscope is used to read the ac voltages, convert the peak-to-peak values to rms values before making any calculations.

1. Amplifier input ac current, i_{in}.

$$i_{in} = \frac{v_{in} - v_{be}}{1000} = ?$$

2. Amplifier input resistance, R_{in}.

$$R_{in} = \frac{v_{be}}{i_{in}} = ?$$

3. Ac output current, i_o.

$$i_o = \frac{v_{ce}}{R_C} = ?$$

4. Voltage gain of the amplifier, A_v.

$$A_v = \frac{v_{ce}}{v_{be}} = ?$$

5. Current gain of the amplifier, A_i.

$$A_i = \frac{i_o}{i_{in}} = ?$$

DISCUSSION QUESTIONS

1. If it is assumed that V_{CG} = 8 V in step B.2. answer and discuss the following questions:

   a. Why should V_{EG} be approximately equal to 7 V?

   b. Find the values of I_C and I_E.

   c. What must be done with the voltage divider circuit of R_1 and R_2 to change I_C to 5 mA?

2. If a voltage divider with an emitter resistor R_E is used in a circuit, then the biasing for the CE amplifier circuit is β independent. In other words, if β changes, I_C and V_{CE} remain relatively constant. Discuss this circuit characteristic, using the values determined in step A of the Procedure to substantiate the discussion.

3. Compare the theoretical and actual test results listed in Table 7-3. Discuss and give reasons for any discrepancies. From the data, what is the approximate β_{dc} of the transistor?

4. Compare the theoretical and actual test results listed in Table 7-4. Discuss and give reasons for any discrepancies. From the data, what is the approximate β_{dc} of the transistor?
5. The power gain of the amplifier is determined by multiplying the voltage gain by the current gain. What is the power gain of the CE amplifier circuit? Is the power gain larger than the voltage gain? Why? Compare the power gain of the CE amplifier to the power gain of the CB amplifier. Discuss the comparison.
6. List some general conclusions about the CE amplifier. For what applications can it be used?

LABORATORY EXERCISE 7-3: COMMON-COLLECTOR (EMITTER FOLLOWER) AMPLIFIER

PURPOSE

- To predict the dc and ac performance of an emitter follower amplifier.
- To construct and then observe the operation of an emitter follower amplifier.
- To measure the dc operating point of an emitter follower amplifier.
- To measure the ac performance quantities of an emitter follower amplifier, including the input and the output resistance and voltage and current gain.

MATERIALS

1 NPN transistor, 2N4425, 2N3405 or equivalent
1 Dc power supply, 9 V
1 VTVM, or solid-state voltmeter
1 Ac VTVM, use an oscilloscope if an ac VTVM is not available
1 Signal generator
1 Oscilloscope
2 Electrolytic capacitors, 25 μ F, 50 V
1 Resistor, 1 k Ω, ½ W
1 Resistor, 10 k Ω, ½ W
1 Resistor, 22 k Ω, ½ W
1 Resistor, 82 k Ω, ½ W
1 Linear potentiometer, 500 Ω, 2 W
1 Switch, SPST to connect the power supply to the circuit

PROCEDURE

A. 1. Convert figure 7-14, the experimental emitter follower or common-collector amplifier, to figure 7-10, the typical circuit. The method used is the same as that used for the CE amplifier.

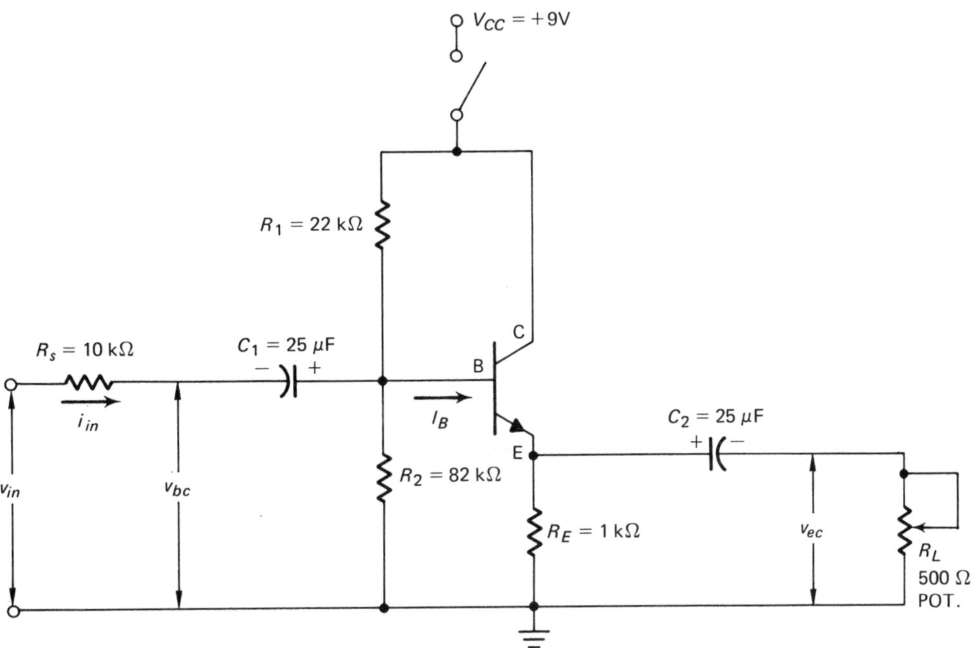

FIG. 7-14 EXPERIMENTAL EMITTER FOLLOWER AMPLIFIER CIRCUIT

$$V_{BB} = \left[\frac{82\,000}{22\,000 + 82\,000}\right] 9\text{ V} = ?$$

$$R_B = \frac{22\,000 \times 82\,000}{22\,000 + 82\,000} = ?$$

Redraw the circuit.

2. Perform the following calculations for each value of β given in Tables 7-6 and 7-7, page 128. Record the calculated values in Table 7-6. Assume that a silicon NPN transistor is used and calculate the dc base current for the experimental emitter follower amplifier. For a silicon transistor, the voltage V_{BE} = 0.6 V.

$$I_B = \frac{V_{BB} - V_{BE}}{R_B + (\beta + 1)\ R_E}$$

$$\beta = 300, I_B = ?$$

$$\beta = 200, I_B = ?$$

$$\beta = 100, I_B = ?$$

3. Calculate the emitter current I_E for each value of I_B in step A.2. Record the values in Table 7-6.

$$I_E = (\beta + 1)\ I_B$$

4. Calculate voltages V_{EG} and V_{CE} for the currents calculated in step A.3. Record the values in Table 7-6.

$$V_{EG} = I_E R_E$$

$$V_{CE} = V_{CC} - V_{EG}$$

TABLE 7-6 Dc (biasing) Conditions

	β	I_B in μA	I_E in mA	V_{BG} in V	V_{EG} in V	$V_{CE} = V_{CC} - V_{EG}$ in V
Theoretical Results	300			–		
	200			–		
	100			–		
Actual Test Results	–	–				

5. Determine the input resistance, r_{ic}, of the experimental amplifier circuit for each value of β, using the following equations. Record the values in Table 7-7.

$$r_{ib} = \frac{50\ \text{mV}}{I_E}$$

$$r_{ic} = (\beta + 1)\ (r_{ib} + R_E)$$

6. Find the amplifier input resistance R_{in}, for each value of r_{ic} determined in step A.5. Record the values in Table 7-7.

$$R_{in} = r_{ic} \parallel R_B$$

7. Find the approximate voltage gain for the amplifier, using the following equation. Record the values in Table 7-7.

$$A_{v\,(CC)} = \frac{R_E}{r_{ib} + R_E}$$

TABLE 7-7 Ac Performance Quantities

	β	r_{ic} in Ω	i_{in} in mA	R_{in} in Ω	i_o in mA	R_o in Ω	$A_{v(CC)}$	$A_{i(CC)}$
Theoretical Results	300		–		–	–		
	200		–		–	–		
	100		–		–	–		
Actual Test Results	–	–						

8. Find the approximate current gain for this amplifier. The current gain of the transistor is approximately equal to β_{ac} (beta). However the voltage divider drains off a considerable amount of the input current. The base current to the transistor may be found by applying the current divider principle, as shown below.

$$i_b = i_{in} = \frac{R_B}{R_B + r_{ic}}$$

$$R_B = R_1 \parallel R_2$$

$$A_{i\,(CC)} \cong \left[\frac{R_B}{R_B + r_{ic}}\right] \beta$$

B. Connect the circuit as shown in figure 7-14, but omit resistor R_S and potentiometer R_L. These resistors will be inserted in later steps. With R_S removed, $v_{in} = v_{bc}$.

C. 1. To check the operating point of the circuit, measure the dc voltage V_{EG}. The voltage V_{EG} should be between 3 and 6 volts. If this voltage range cannot be obtained, change the biasing resistors in the voltage divider circuit. Try changing R_2 first.

2. When the voltage is within the desired range, complete the Actual Test Results row of Table 7-6. The current I_E is calculated using the dc voltages measured here and the formulas given in step A. If the biasing resistors R_1 or R_2 were changed, indicate the correct values in the circuit in figure 7-14.

D. 1. Connect an oscilloscope across the output, v_{ec}.

2. Apply a 1-kHz input signal from a sine-wave generator.

3. Increase the input signal until the output sine wave distorts. Then decrease the input until a nondistorted sine wave appears across the output. Since the input signal will be less than 100 mV, a voltage divider may be required across the signal generator.

4. Measure the input voltage, v_{bc}, and output voltage, v_{ec}, with an ac VTVM. Record these values in Table 7-8. If an ac VTVM is not available, use an oscilloscope and record the peak-to-peak readings in Table 7-8.

E. Decrease the input signal to the smallest readable value (approximately 1 mV) and record the input and output voltages in Table 7-8.

TABLE 7-8 Ac Circuit Readings

Step	v_{in} in V	v_{bc} in V	v_{ec} in V	$R_L = R_o$ in Ω	A_v Calculated
D	—			—	
E	—			—	
F				—	—
G					—

F. 1. To determine the input current and input resistance of the amplifier circuit, insert resistor $R_S = 10\ k\ \Omega$ before the capacitor as shown in figure 7-14.

2. Increase the input signal so that the voltage at v_{bc} is the same as its value in step D.4.

3. Read and record v_{in}, v_{bc}, and v_{ec} in Table 7-8.

G. 1. To measure the output resistance of the amplifier circuit, connect the potentiometer R_L as shown in figure 7-14.

2. Adjust R_L until the output voltage, v_{ce}, is one-half of its value in step F.2. Record the voltage readings in Table 7-8.

3. Remove R_L from the circuit without changing the setting. Measure R_L and record this value as R_o in Tables 7-7 and 7-8.

H. Calculate the following ac performance quantities using the values recorded in step F of Table 7-8. List the results in Table 7-7. If an oscilloscope is used to read the ac voltages, convert the peak-to-peak values to rms values before making any calculations.

1. Amplifier input ac current, i_{in}.

$$i_{in} = \frac{v_{in} - v_{bc}}{1000} = ?$$

2. Amplifier input resistance, R_{in}.

$$R_{in} = \frac{v_{bc}}{i_{in}} = ?$$

3. Ac output current, i_o.

$$i_o = \frac{v_{ec}}{R_E} = ?$$

4. Voltage gain of the amplifier, A_v.

$$A_v = \frac{v_{ec}}{v_{bc}} = ?$$

5. Current gain of the amplifier, A_i.

$$A_i = \frac{i_o}{i_{in}} = ?$$

DISCUSSION QUESTIONS

1. For step C of the Procedure, assume that $V_{EG} = 2$ V. Answer and discuss the following questions.

   a. Find the value of V_{CE}.

   b. Find the value of I_E.

   c. What must be done with the voltage divider circuit of R_1 and R_2 to change I_E to 5 mA?

2. Compare the theoretical and actual test results listed in Table 7-6. Discuss and give reasons for any discrepancies. From the data, what is the approximate β_{dc} of the transistor?

3. In step D.4. of the Procedure, is the output voltage approximately the same as the input voltage? Why? Discuss this conclusion.

4. Compare the theoretical and actual test results listed in Table 7-7. Discuss and give reasons for any discrepancies. From the data, what is the approximate β_{ac} of the transistor?
5. The power gain of the amplifier may be determined by multiplying the voltage gain by the current gain. What is the power gain of this emitter follower amplifier circuit? Is the power gain approximately the same as the current gain? Why? Compare the emitter follower power gain to the power gain for the CB and CE amplifiers. Discuss the comparisons.
6. List some general conclusions about the emitter follower amplifier. For what applications can it be used?
7. In Table 7-9, list the quantities from the actual test results in the various tables for the CB, CE, and CC amplifier circuits. Compare these amplifier circuits. Which amplifiers have the highest and lowest values of R_{in}, R_o, A_v, and A_i?

TABLE 7-9 Comparison of CB, CE, and CC Amplifiers

Type of Amplifier	R_{in} in Ω	R_o in Ω	A_v	A_i
CB				
CE				
CC (Emitter Follower)				

EXTENDED STUDY TOPICS

1. Using figure 7-3b, determine the ac current gain at $V_{CE} = 7$ V with an I_B change of 0.4 to 0.6 mA. Find the emitter current change for these conditions if $I_{CEO} = 10\ \mu$A.
2. Find the missing quantities in the table below.

α	β	I_C in mA	I_B in mA	I_E in mA	I_{CBO} in μA	I_{CEO} in μ A
a.			1.0	10	10	
b.	50		0.5			10
c.			2.5	70		0
d. 0.95				8	5	

3. For the circuit of figure 7-6, assume that $R_C = 10$ k Ω and $V_{CC} = 10$ V. Find the collector saturation current. If $\beta = 100$, what is the base current if $I_{CBO} = 0$?
4. A CE transistor amplifier experiences a change in V_{BE} of 1.5 mV, a change in V_{CE} of 12V, a change in the base current of 2 μA, and a change in the collector current of 100 μA. Find r_{ie}, r_{oe}, and the voltage gain if $R_C = 1$ kΩ.
5. For the circuit in figure 7-11, $r_{ie} = 600$ Ω and $\beta = 75$. Find r_{ic}.
6. Calculate the voltage gain for Topic 5.

Transistor specifications and graphical analysis

OBJECTIVES

After studying this unit, the student will be able to discuss and demonstrate an understanding of the basic principles of:

- The interpretation of transistor data sheets.
- Derating a transistor when it is operated at elevated temperatures.
- Constructing power dissipation curves.

TRANSISTOR SPECIFICATION SHEETS AND TERMINOLOGY

Figure 8-1 and figure 8-2 are reproductions of complete manufacturer transistor specification sheets. Such specifications provide considerable information regarding a particular transistor or family of transistors. However, the majority of manufacturers' specification sheets are not as complete as those shown. As a result, it may be necessary to use curve tracers or other tests to obtain the desired parameters of a particular transistor.

The transistors covered by the specification sheets in figures 8-1 and 8-2 may be used for linear medium power amplifiers and medium speed switches. The 2N525 transistor is a germanium PNP transistor, and the 2N4425 transistor is a plastic-encased silicon NPN transistor. These transistors are used in many of the problems and in the amplifier circuits presented in this unit. The techniques of using the transistors, however, apply to all bipolar junction transistors, provided the proper parameters and power supplies are used.

It is recommended that the transistor specification sheets be studied carefully. The electronics technician will often be required to read, comprehend, and interpret these specifications quickly and accurately, when repairing or designing circuits. The balance of this unit is designed to familiarize the student with the correct use of these specification sheets.

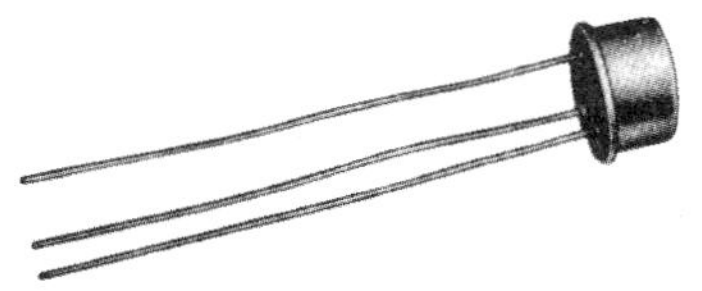

INDUSTRIAL TRANSISTORS

The General Electric types 2N524, 2N525, 2N526, and 2N527 are PNP germanium alloy transistors recommended for extremely reliable medium power amplification and switching in low frequency military and industrial applications.
Maximum operational reliability is assured by inclusion of military environmental testing. Each transistor is stabilized for at least 100 hours at 100°C minimum and is subjected to a hermetic seal test insuring a maximum leak rate of 10^{-10} cc/sec.
Reliable circuit design is assured by a unique Reliability-Index, and high temperature I_{CBO} and low temperature forward current gain 1000 hour life end points.
The high value of Reliability-Index is achieved by exacting control of parts, processes, and getter encapsulation to prevent junction contamination.

absolute maximum ratings (25°C)

Voltages		
Collector to Base	V_{CBO}	− 45 volts
Collector to Emitter ($R_{BE} \leqq 10K$)	V_{CER}	− 30 volts
Emitter to Base	V_{EBO}	− 15 volts
Collector Current	I_C	− 500 ma
Temperatures		
Storage	T_{STG}	− 65 to + 100°C
Operating	T_J	− 65 to + 85°C
Lead Temperature, 1/16″ ± 1/32″ from case for 10 seconds maximum	T_L	260°C
Total Transistor Dissipation		
Free Air (Derate 3.75 mw/°C increase in ambient above 25°C)	P_T	225 mw
Clip-on Heat Sink in Free Air (Derate 5.0 mw/°C)	P_T	300 mw
Infinite Heat Sink (Derate 9.15 mw/°C)	P_T	550 mw

DIMENSIONS WITHIN JEDEC OUTLINE TO-5

NOTE 1: This zone is controlled for automatic handling. The variation in actual diameter within this zone shall not exceed .010.

NOTE 2: Measured from max. diameter of the actual device.

NOTE 3: The specified lead diameter applies in the zone between .050 and .250 from the base seat. Between .250 and .5 maximum of .021 diameter is held. Outside of these zones the lead diameter is not controlled. Leads may be inserted, without damage, in .031 holes while transistor enters .371 hole concentric with lead hole circle.

APPROX WEIGHT: .05 OZ
ALL DIMENSIONS IN INCHES

.370 MAX .360 MIN
.335 MAX .325 MIN
.100 MIN (NOTE 1)
.260 MAX .215 MIN
1.5 MIN
3 LEADS .017 +.002 −.001 (NOTE 3)
.210 MAX .190 MIN
THIS LEAD GROUNDED TO HOUSING
2 B
1 E
3 C
45° ± 3°
.040 MAX .029 MIN (NOTE 2)
90°
.031 ± .003

reliability-index

The Reliability-Index (RI_1) has been developed by General Electric to increase customer assurance of stable life performance. It is based on quality control information on each lot for normalized distribution and dispersion shift of the gain characteristic. A factor of 4.0 or greater for RI indicates outstanding extended life performance.

The Reliability-Index is obtained by:
1. Computing the percentage shift (+ or −) in the forward current gain of each unit on life.
2. Determining the 10th, 50th and 90th percentiles in a distribution of the individual unit percent shifts.
3. Adding the magnitude of the 50th percentile to the magnitude of the algebraic difference between the 90th and 10th percentiles, and
4. Multiplying the reciprocal by 100.

expressed algebraically:

$$RI_1 = \frac{100}{|\alpha_{50}| + |\alpha_{90} - \alpha_{10}|}$$

Where α_{50}, α_{90}, and α_{10} are the particular percentile values of a distribution of α_i, and

$$\alpha_i = \frac{h_{FE}F_i - h_{FE}I_i}{h_{FE}I_i}$$

Where $h_{FE}F_i$ is the final and $h_{FE}I_i$ the initial value of the forward current gain of the ith transistor.

FIG. 8-1

ELECTRICAL CHARACTERISTICS (T_A=25°C)

TEST	CONDITIONS	SYMBOL	UNITS	2N524 MIN.	2N524 TYP.	2N524 MAX.
DC Characteristics						
Subgroup I						
Collector to Base Voltage	I_C = −200 μa	V_{CBO}	volts	− 45		
Collector to Emitter Voltage	R_{BE} = 10K ohms, I_C = −600 μa	V_{CER}	volts	− 30		
Reach-through Voltage		V_{RT}	volts	− 30		
Emitter Cutoff Current	V_{EB} = −15v	I_{EBO}	μa		− 3	− 10
Subgroup II						
Collector-Cutoff Current	V_{CB} = −30v	I_{CBO}	μa		− 3	− 10
Forward Current Transfer Ratio	I_C = −20 ma, V_{CE} = −1v (Note 1)	h_{FE}		25(28)	35	(38)42
Forward Current Transfer Ratio	I_C = −100 ma, V_{CE} = −1v	h_{FE}		23	30	
Subgroup III						
Base Input Voltage, Common Emitter	I_C = −20 ma, V_{CE} = −1v	V_{BE}	volts	−.220	−.250	−.320
Collector Saturation Voltage	I_C = −20 ma, I_B as shown	V_{CE}(SAT)	volts		−.075	−.130
		@I_B	ma		−2.00	
Small Signal Characteristics						
Subgroup I						
AC Forward Current Transfer Ratio	I_E = 1 ma, V_{CE} = −5v, f = 1kc	h_{fe}		18	30	41
Input Impedance	I_E = 1 ma, V_{CB} = −5v, f = 1kc	h_{ib}	ohms	26	30	34
Output Admittance	I_E = 1 ma, V_{CB} = −5v, f = 1kc	h_{ob}	μmhos	0.1	0.7	1.3
Reverse Voltage Transfer Ratio	I_E = 1 ma, V_{CB} = −5v, f = 1kc	h_{rb}	x10^{-4}	1	4	10
Subgroup II						
Cutoff Frequency	I_E = 1 ma, V_{CB} = −5v	f_{hfb}	mc	0.8	2.5	5.0
Output Capacity	I_E = 1 ma, V_{CB} = −5v, f = 1mc (Note 1)	C_{ob}	pf	5	18	(30)40
Noise Figure	I_E = 1 ma, V_{CB} = −5v, f = 1kc, BW = 1 cps, Rg = 1000 ohms (Note 1)	NF	db	(1)	5	(8)15

TEST	2N525 MIN.	2N525 TYP.	2N525 MAX.	2N526 MIN.	2N526 TYP.	2N526 MAX.	2N527 MIN.	2N527 TYP.	2N527 MAX.	% AQL	INSP. LEVEL
Collector to Base Voltage	− 45			− 45			− 45				
Collector to Emitter Voltage	− 30			− 30			− 30				
Reach-through Voltage	− 30			− 30			− 30			0.65	II
Emitter Cutoff Current		− 3	− 10		− 3	− 10		− 3	− 10		
Collector-Cutoff Current		− 3	− 10		− 3	− 10		− 3	− 10		
Forward Current Transfer Ratio (I_C = −20 ma)	34(38)	50	(59)65	53(59)	70	(80)90	72(80)	95	(110)121	0.65	II
Forward Current Transfer Ratio (I_C = −100 ma)	30	45		47	60		65	80			
Base Input Voltage, Common Emitter	−.200	−.235	−.300	−.190	−.230	−.280	−.180	−.225	−.260		
Collector Saturation Voltage		−.080	−.130		−.085	−.130		−.090	−.130	1.0	L6
@I_B		−1.33			−1.00			−0.67			
AC Forward Current Transfer Ratio	30	44	64	44	64	88	60	80	120		
Input Impedance	26	29	33	26	28	32	26	28	31		
Output Admittance	0.1	0.6	1.2	0.1	0.5	1.0	0.1	0.4	0.9	1.0	L6
Reverse Voltage Transfer Ratio	1	5	11	1	6	12	1	7	14		
Cutoff Frequency	1.0	3.0	5.5	1.3	3.5	6.5	1.5	4.0	7.0		
Output Capacity	5	18	(30)40	5	18	(30)40	5	18	(30)40		
Noise Figure	(1)	4	(7)15	(1)	3	(6)15	(1)	2	(5)15		

Note 1: A minimum of 95% of the distribution is normally contained between values in parenthesis.

FIG. 8-1 (continued)

RELIABILITY SPECIFICATIONS

TEST	CONDITIONS	AQL	SAMPLING LEVEL
Subgroup I	MIL-S-19500		
Physical Inspection		2.5	L5
Appearance		1.0	L8
Lead Solderability		1.0	L8
Subgroup II	MIL-S-19500		
Temperature Cycling	−65° to 100°C, 10 cycles	4.0	L5
Thermal Shock	0° to 85°C, 5 cycles		
Moisture Resistance			
Subgroup III	MIL-S-19500		
Shock, operating	1000 G, 5 blows each orientation of approximately 0.8 msec.	4.0	L5
Constant Acceleration	10,000 G		
Vibration Fatigue	10 G		
Vibration, Variable Frequency	20 G		
Subgroup IV	MIL-S-19500		
Lead Fatigue		4.0	L5

END POINTS (Subgroup II and III)	CONDITIONS	Symbol	Units	Min.	Max.
Collector Cutoff Current	$V_{CB} = -30v, T_A = 25°C$	I_{CBO}	μa		−15
Emitter Cutoff Current	$V_{EB} = -15v, T_A = 25°C$	I_{EBO}	μa		−15
Forward Current Transfer Ratio	$I_C = -20ma, V_{CE} = -1V, T_A = 25°C$				
2N524		h_{FE}		22	46
2N525		h_{FE}		30	71
2N526		h_{FE}		48	100
2N527		h_{FE}		65	133

TEST	CONDITIONS	MAXIMUM FAILURE RATE, λ (in percent per 1000 hours with 90% confidence)	
Subgroup V			
Storage Life (1000 hours)	MIL-S-19500 Method B 100° + 10°C, −0°C	5	
Subgroup VI			
Intermittent Power Life (1000 hours)	MIL-S-19500 Method B 225 mw, $T_A = 25° \pm 4°C$ $V_{CB} = -22.5v, I_E = -10$ ma On 50 ± 2 min, Off 10 ± 2 min	5 10	(25°C End Points) (−55°C and +70°C End Points)

TEST	CONDITIONS	Symbol	Units	Min.	Typ.	Max.
END POINTS (Subgroup VI)						
Reliability-Index	Based on 25°C Forward Current Transfer Ratio, Subgroup VI	RI_1		4.0	12	
Collector Cutoff Current	$V_{CB} = -10v, T_A = +70° \pm 1°C$	I_{CBO}	μa		−80	−150
Forward Current Transfer Ratio	$I_C = -20$ ma, $V_{CE} = -1v$, $T_A = -55° \pm 2°C$					
2N524		h_{FE}		11	23	
2N525		h_{FE}		15	33	
2N526		h_{FE}		24	46	
2N527		h_{FE}		32	63	
END POINTS (Subgroup V and VI)						
Collector Cutoff Current	$V_{CB} = -30v, T_A = 25°C$	I_{CBO}				−15
Forward Current Transfer Ratio	$I_C = -20$ ma, $V_{CE} = -1v$, $T_A = 25°C$					
2N524		h_{FE}	μa	22		46
2N525		h_{FE}		30		71
2N526		h_{FE}		48		100
2N527		h_{FE}		65		133

FIG. 8-1 (continued)

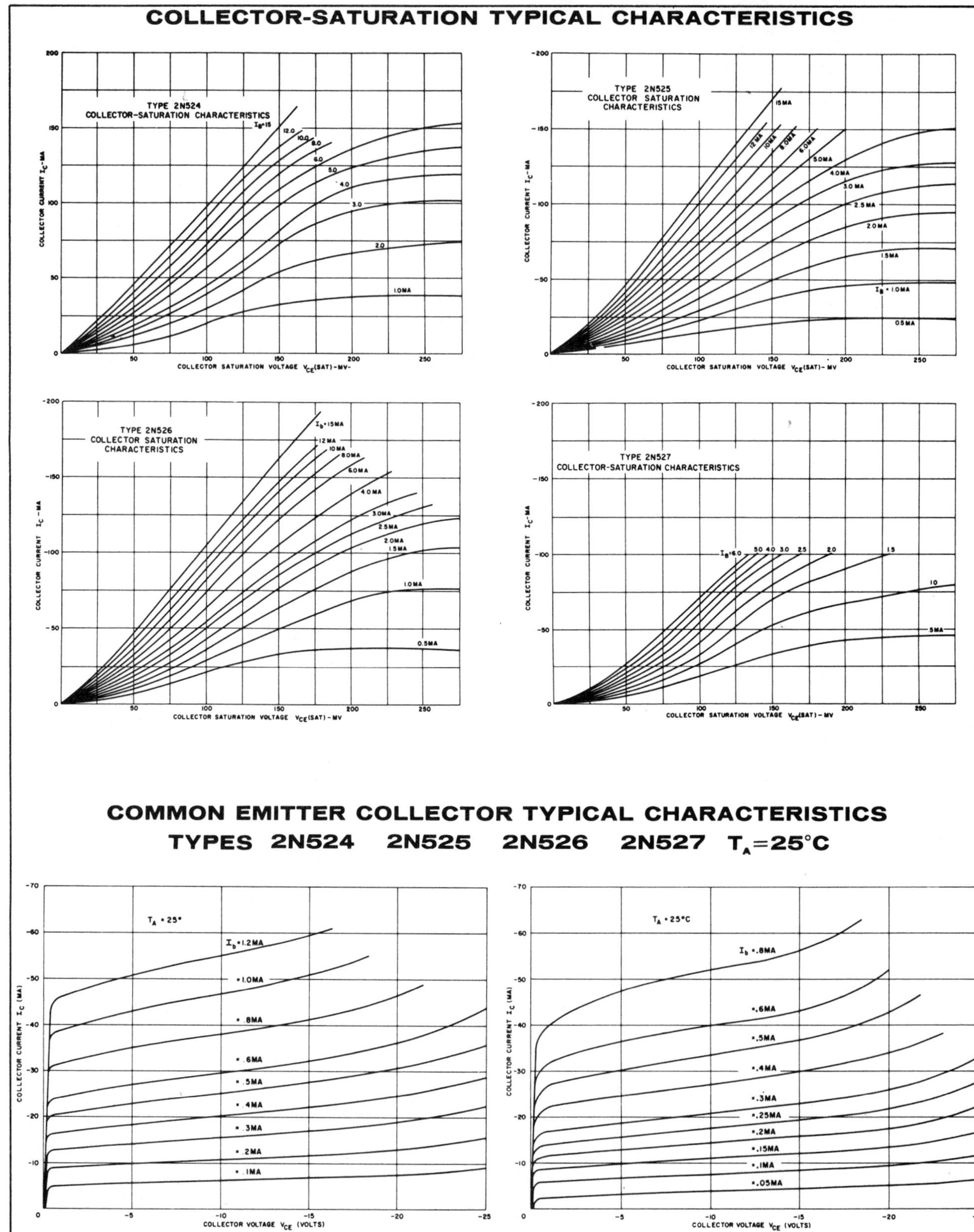

FIG. 8-1 (continued)

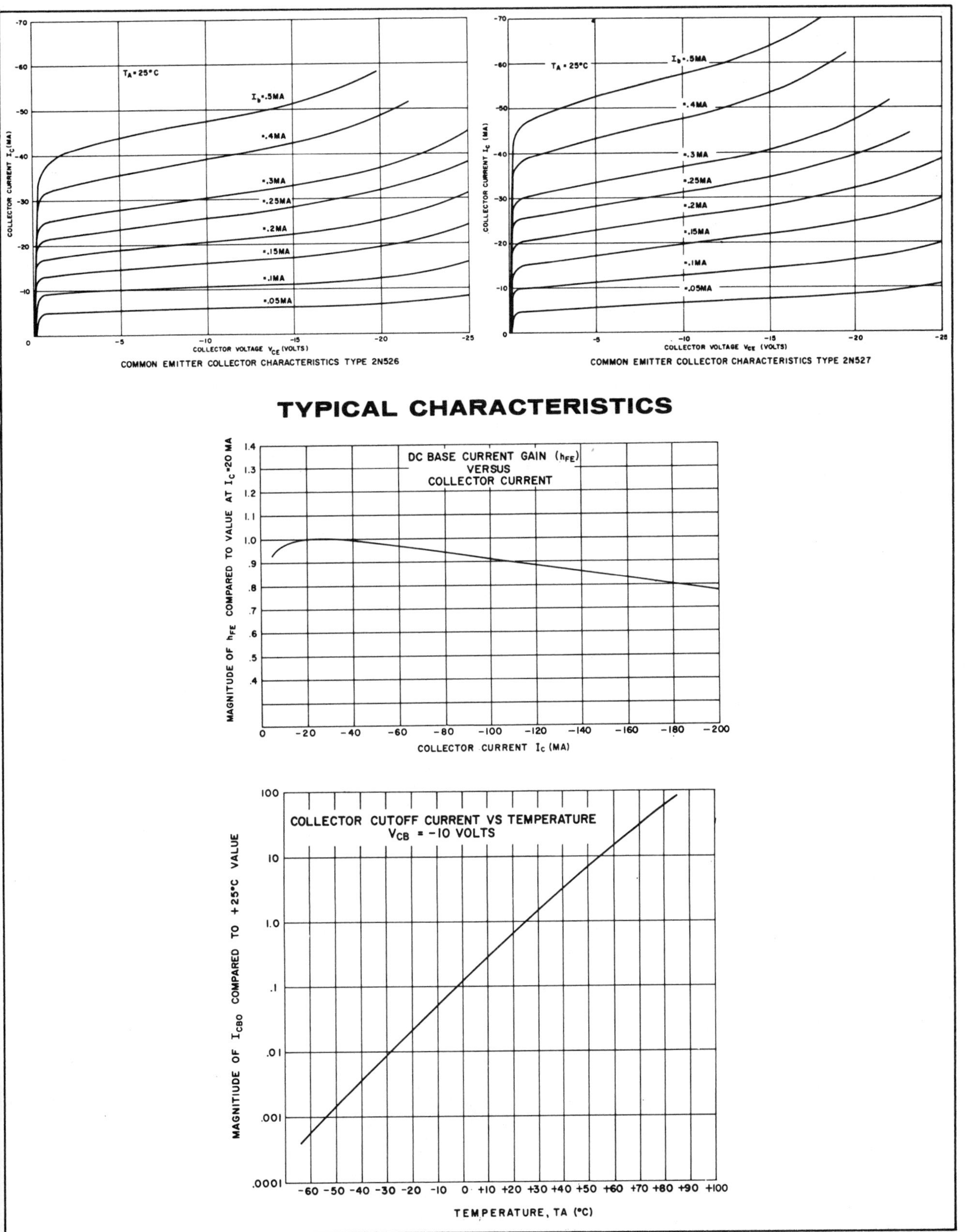

FIG. 8-1 (continued)

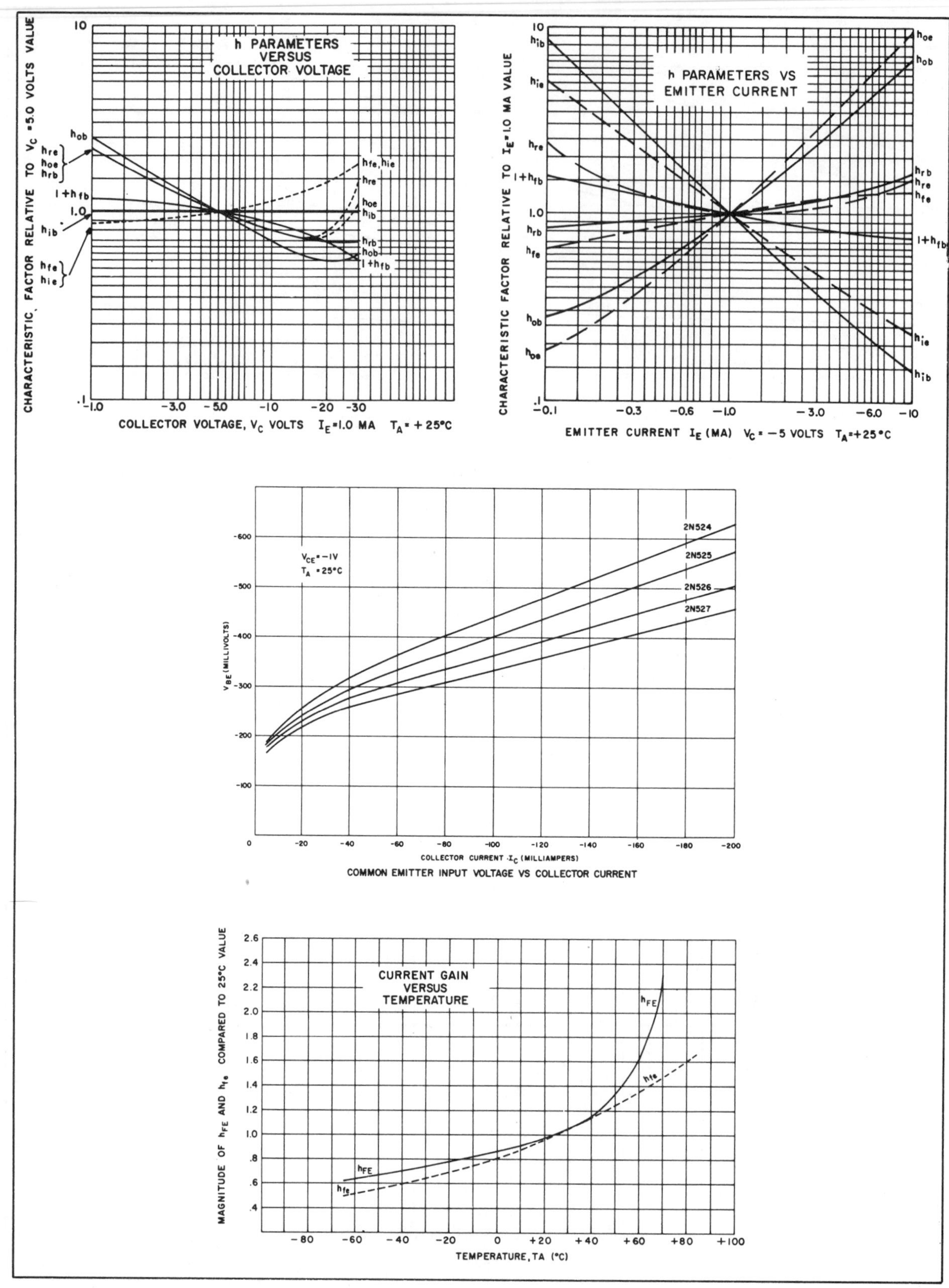

FIG. 8-1 (continued)

Silicon Consumer — Industrial Transistors
PLANAR EPITAXIAL PASSIVATED

NPN

2N4424 2N4425

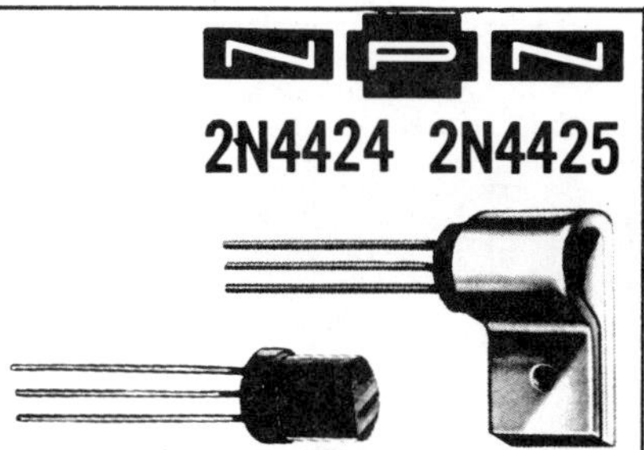

The General Electric 2N4424 and 2N4425 types are NPN, silicon, planar, passivated, epitaxial transistors intended for general purpose industrial circuits. These transistors are especially suited for high level linear amplifiers or medium speed switching circuits in industrial control applications.

FEATURES:

Low Saturation Voltage
High Beta
900 mW @ 25°C Case — 2N4425
360 mW @ 25°C Free Air — 2N4424

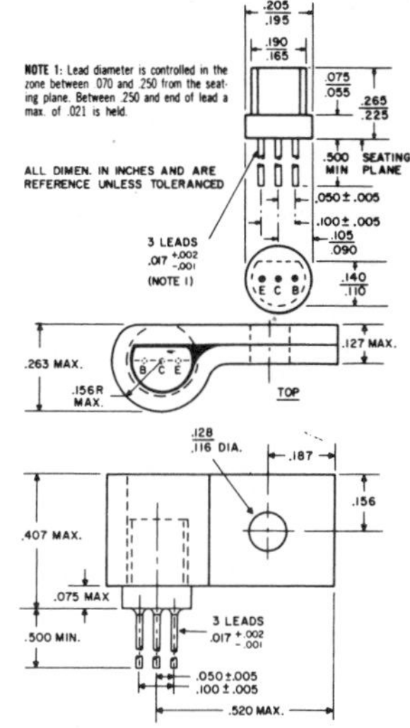

absolute maximum ratings: (25°C) (unless otherwise specified)

		2N4424	2N4425	
Voltages				
Collector to Emitter	V_{CEO}	40	40	V
Emitter to Base	V_{EBO}	5	5	V
Collector to Base	V_{CBO}	60	60	V
Current				
Collector (Steady State)*	I_C	500	500	mA
Dissipation				
Total Power (Free Air at 25°C)**	P_T	360	560	mW
Total Power (Free Air at 65°C)**	P_T	250	380	mW
Total Power (Heatsink at 25°C)***	P_T	—	900	mW
Temperature				
Storage	T_{stg}	−55 to +150		°C
Operating	T_J		+150	°C
Lead soldering, 1/16″ ± 1/32″ from case for 10 sec. max.	T_L		+260	°C

*Determined from power limitations due to saturation voltage at this current.
**Derate 2.88mW/°C increase in ambient temperature above 25°C.
***Derate 7.2 mW/°C for rise in heatsink temperature above 25°C.

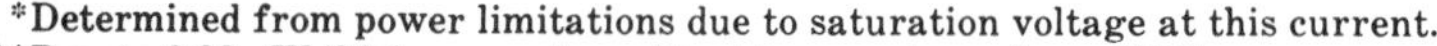

electrical characteristics: (25°C) (unless otherwise specified)

DC CHARACTERISTICS

		Min.	Max.	
Collector Cutoff Current (V_{CB} = 40V)	I_{CBO}		30	nA
(V_{CB} = 40V, T_A = 100°C)	I_{CBO}		10	μA
(V_{CB} = 40V)	I_{CES}		30	nA
Emitter Cutoff Current (V_{EB} = 5V)	I_{EBO}		100	nA
Forward Current Transfer Ratio (V_{CE} = 4.5V, I_C = 2 mA)	h_{FE}	180	540	
Collector Emitter Breakdown Voltage (I_C = 10 mA)	$V_{(BR)CEO}$	40		V
Collector Base Breakdown Voltage (I_C = 10 μA)	$V_{(BR)CBO}$	60		V
Emitter Base Breakdown Voltage (I_E = 0.1 μA)	$V_{(BR)EBO}$	5		V
Collector Saturation Voltage (I_B = 3 mA, I_C = 50 mA)	$V_{CE(sat)}$		.30	V
Base Saturation Voltage (I_B = 3 mA, I_C = 50 mA)	$V_{BE(sat)}$		.85	V

SMALL SIGNAL CHARACTERISTICS

		Min.		
Forward Current Transfer Ratio Collector Voltage (V_C = 4.5V, I_C = 2 mA, f = 1 kHz)	h_{fe}	180		

			Typical	
Forward Current Transfer Ratio	(V_{CE} = 10V, I_C = 1 mA, f = 1 kHz, T_A = 25°C)	h_{fe}	180	
Input Impedance		h_{ie}	5100	ohms
Output Admittance		h_{oe}	14	μmhos
Voltage Feedback Ratio		h_{re}	.27	$\times 10^{-3}$

FIG. 8-2

SPECIFICATIONS ON PAGES 139-142 COURTESY OF GENERAL ELECTRIC SEMICONDUCTOR PRODUCTS DEPARTMENT, SYRACUSE, N.Y.

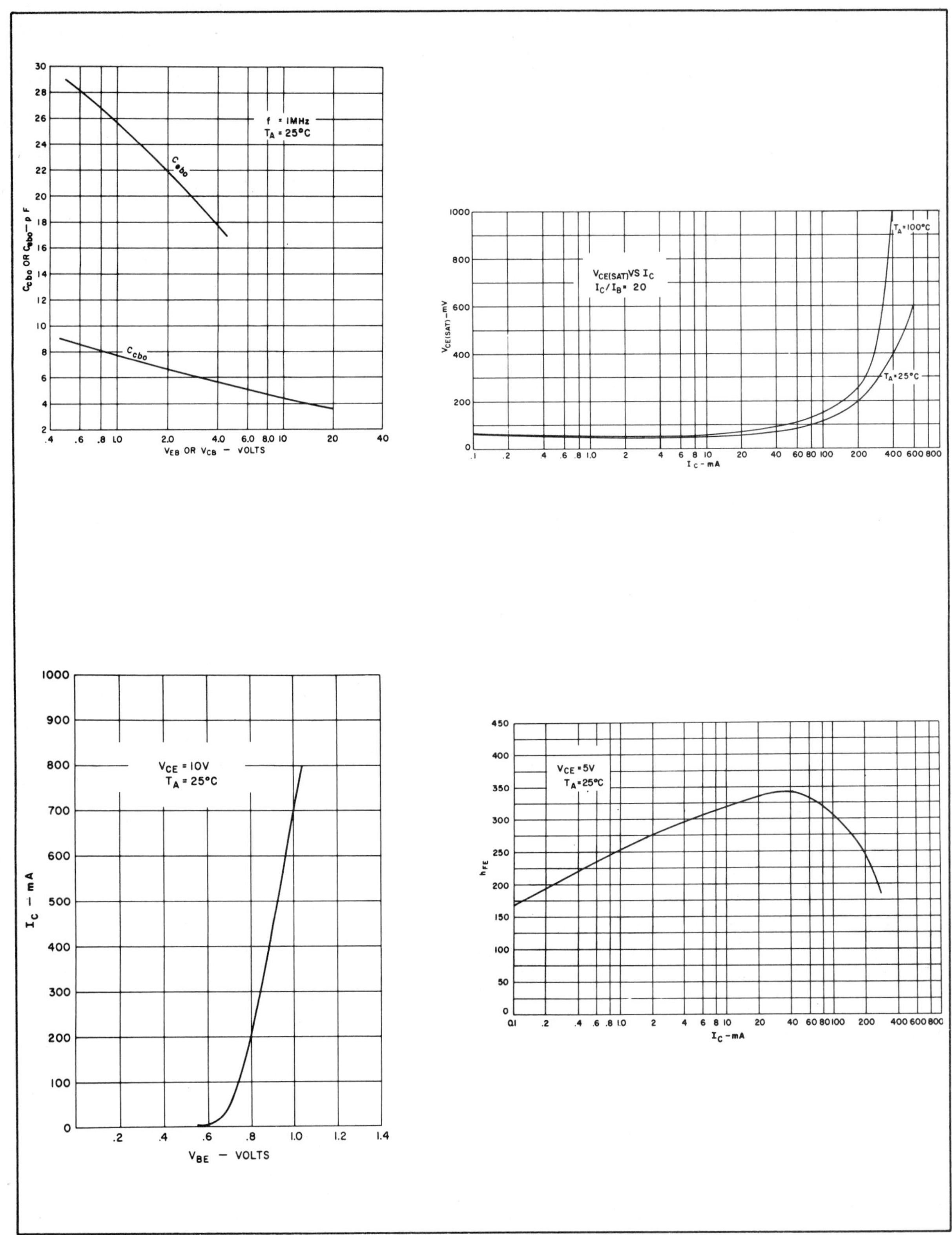

FIG. 8-2 (continued)

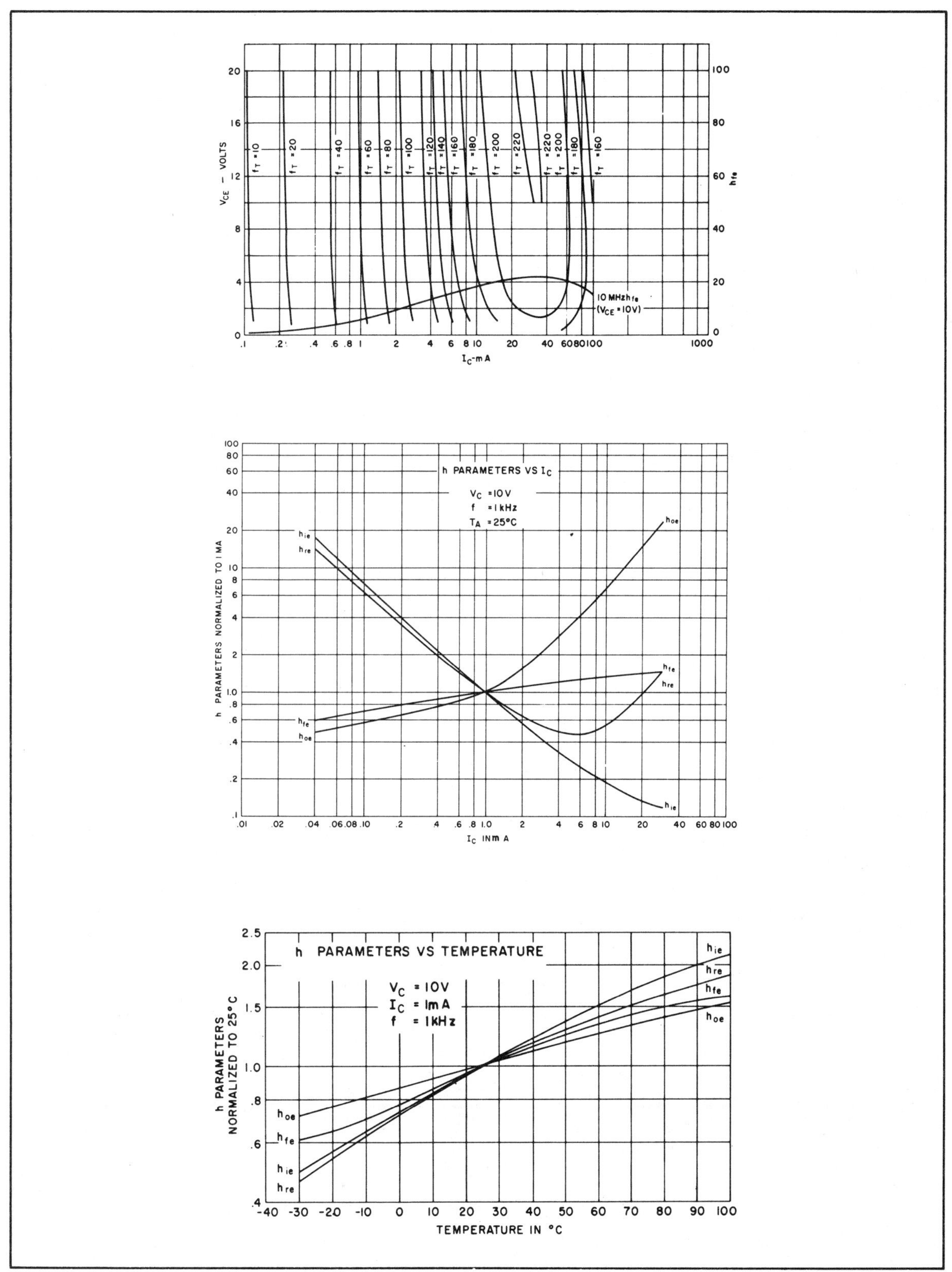

FIG. 8-2 (continued)

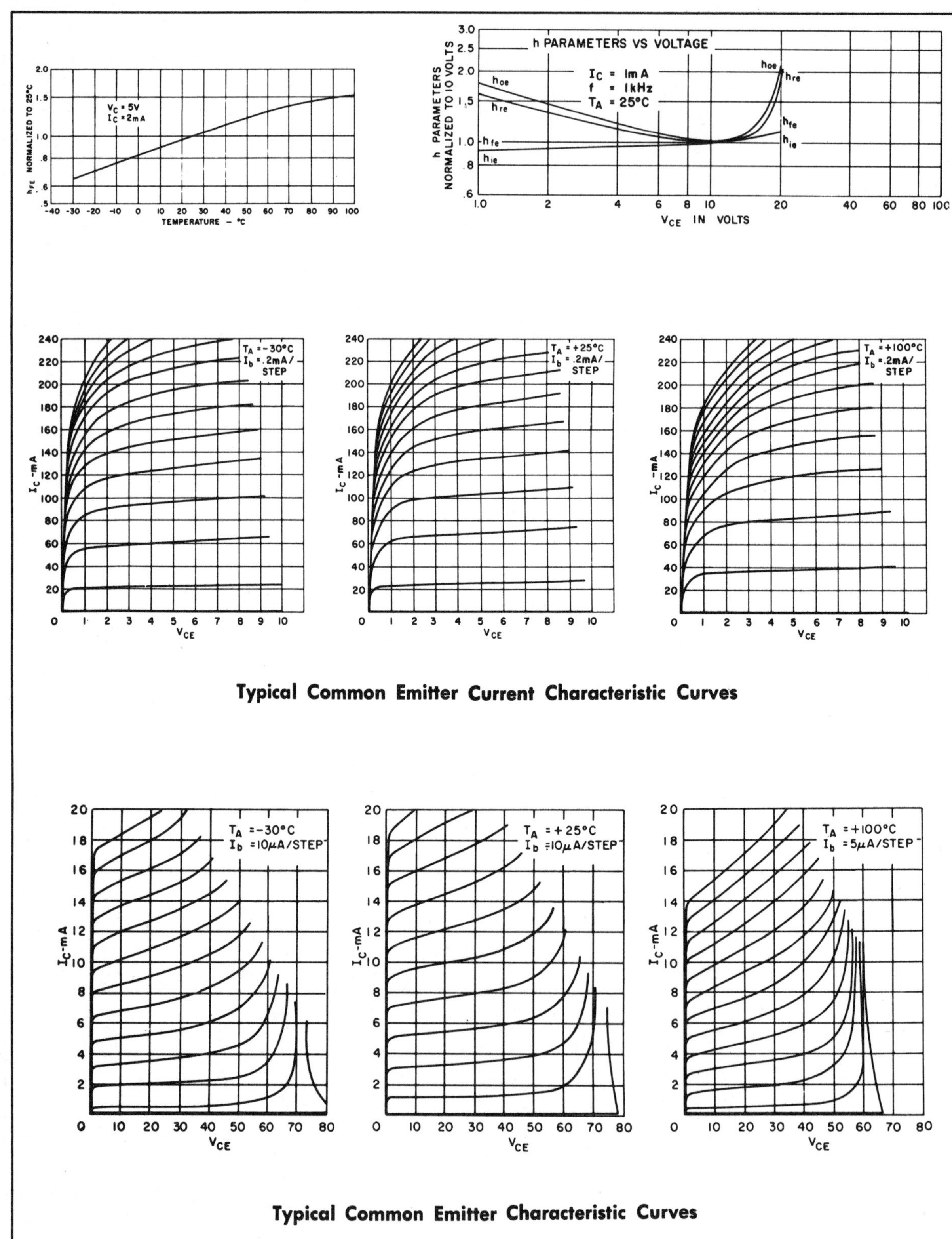

Typical Common Emitter Current Characteristic Curves

Typical Common Emitter Characteristic Curves

FIG. 8-2 (continued)

Transistor specification sheets are divided into four general categories of information:

- General Description
- Absolute Maximum Ratings
- Electrical Characteristics at 25°C
- Graphs of Transistor Characteristics and Parameter Corrections

General Description

The opening paragraph usually contains descriptive information including the transistor *type*, method of *construction*, and *applications.*

Absolute Maximum Ratings at 25°C

This means just what it says, *absolute maximum transistor ratings.* If any of these ratings are *exceeded,* the transistor may be *destroyed,* or it may undergo a change in its characteristics. A transistor cannot heal itself. Note that these ratings are given at 25°C. If the transistor is operated at higher temperatures, *derating factors* must be applied to these values.

Some of the dc symbols found in this section of the specification sheet are defined as follows:

V_{EBO} = reverse-biased emitter-base voltage with the collector open.

V_{CBO} = reverse-biased collector-base voltage with the emitter open.

V_{CEO} = reverse-biased collector-emitter voltage with the base open.

If a B precedes any of these symbols, such as BV_{CBO}, it specifies the dc *breakdown* voltage at a *particular current value.* The dc breakdown voltages are usually given in the electrical characteristics section of the specification sheet.

The symbol T followed by a subscript specifies particular maximum temperatures. For example, T_{STG} is the *storage* temperature, T_J is the junction operating temperature, and T_A is the *ambient* operating temperature.

The power rating of a transistor is of particular interest. The power rating may be given as P_T, *total transistor dissipation,* or P_C, *collector dissipation.* This value represents the power dissipation that the transistor (or its collector) can handle without being destroyed. Eq. 8.1 and 8.2 are used to determine the maximum power dissipation at a particular operating temperature.

$$P_C(T) = P_C(T_B) - X\,(T - T_R) \quad \text{Eq. 8.1}$$

$$P_C(T) = P_C(T_R) - \frac{(T - T_R)}{\theta} \quad \text{Eq. 8.2}$$

where

$P_C(T_R)$ = the maximum collector dissipation at a given reference temperature. This quantity is usually given on the transistor specification sheet at 25°C. The dissipation is expressed in watts (W) or milliwatts (mW).

T_R = The *reference temperature,* usually 25°C.

T = the *operating temperature,* in °C.

X = a *derating factor* that adjusts the transistor dissipation to a changing temperature. This quantity is usually given on the transistor specification sheets. The derating factor is expressed in W/°C, or mW/°C for free air, clip-on heat sinks, or infinite heat sinks.

θ = the *thermal resistance* of the junction to the transistor case, or the junction to free air. The units of the thermal resistance are °C/W, or

°C/mW. This parameter is found by taking the reciprocal of X: $\theta = 1/X$.

$P_C(T)$ = the maximum collector dissipation at a particular operating temperature, in watts (W) or milliwatts (mW).

PROBLEM 1.

A 2N525 transistor is operated in free air at 50°C. Calculate the power dissipated by this transistor at the given operating temperature.

Refer to the specification sheet for the 2N525 transistor to obtain the following parameters.

P_T = 225 mW in free air at 25°C.

X = derating of 3.75 mW/°C increase in ambient above 25°C.

Using Eq. 7.1:

$$P_C(T) = P_C(T_R) - X(T - T_R)$$

$$P_C(50°C) = 225 \text{ mW} - 3.75 \text{ mW/°C} \times (50°C - 25°C)$$

$$= 225 \text{ mW} - 93.7 \text{ mW} = \mathbf{131.3 \text{ mW}}$$

Electrical Characteristics

There are usually several subdivisions in this section of the specification sheet. Depending upon the transistor type and whether the manufacturer provides complete information, this section may contain information on *dc characteristics, small signal characteristics, switching characteristics,* and *high frequency characteristics.* These characteristics will apply only when the transistor is operated at 25° C and at the specified operating point.

The specification sheet always indicates the test conditions under which the parameters were measured. If the design circuit conditions are not the same as the conditions specified, then these values should not be used. An attempt should be made to correct the circuit conditons. If better data cannot be obtained, or you cannot perform your own tests, then use the given values but be aware that any calculations may be in *error* by as much as 100%.

Note that the quantities I_{CBO} or I_{CO} and h_{FE}, which were covered in earlier units of this text, are specified in the dc characteristics section. The parameter h_{FE} or β is useful for switching circuit applications. With the aid of this parameter, the amount of base current, $I_B = I_{CS}/h_{FE}$ can be determined, as well as the collector saturation current, $I_{CS} = V_{CC}/R_C$.

In the small signal characteristics section, the following ac *hybrid* parameters are usually given: h_f, h_i, h_r, and h_o. The letter b or e follows the first subscript letter (f, i, r, or o) to indicate that the parameter measurement was made in the common-base or common-emitter configuration respectively. Note that the h parameters on the specification sheet are given at a particular operating point. If the design circuit is not operating at this particular point, and the given parameters are used, serious errors may occur in the calculations for the circuit. Problems 2 and 3 show how specific parameters are corrected for operating points which differ from those given in the specifications.

If the major transistor application is switching, then the switching characteristics section of the specifications will provide the transient response times for the transistor. The response times are known as t_d, t_r, t_s, and t_f, which are the *delay, rise, storage,* and *fall* times respectively. The response time parameters will not be analyzed until transistor switching applications are covered.

The *high frequency characteristics* section usually contains the following symbols:

f_{hfb} = common-base, small-signal (short circuit), forward-current gain, cutoff frequency. This is the frequency at which the common-base short circuit gain falls 3 dB to 0.707 of its medium value.

C_{ob} = collector-base junction output capacitance

NF = noise figure

Graphs of Transistor Characteristics and Parameter Correction

The graph section of the specification sheet will contain the common-emitter characteristic curves. In addition, the graph section may contain graphs for *parameter corrections* required by changes in the temperature and operating point. Most of these correction curves have *normalized* vertical scales. The term normalized means that the answer obtained from the graph is relative to the value given in the electrical characteristics section at 25°C. Problem 2 and 3 illustrate the use of correction parameters and normalized graphs.

PROBLEM 2.

A 2N525 transistor is operating at 65°C. The normalized I_{CBO} versus temperature curve is shown in figure 8-3. Determine the leakage current I_{CBO} at 65°C.

The vertical scale for I_{CBO} in figure 8-3 is logarithmic and normalized. The 65°C line intersects the curve at the number 21. This means that I_{CBO} at 65°C is 21 times larger than its value at 25°C. The specification sheet for the 2N525 transistor gives a value for I_{CBO} at 25°C of 3 μA. Therefore, the actual I_{CBO} at 65°C is:

$$I_{CBO} = 3\,\mu A \times 21 = \mathbf{63\,\mu A.}$$

If it is necessary to use two normalized graphs to correct the value of a parameter, then the normalized factor is found in each graph. These two values are multiplied and the result is then multiplied by the parameter value given at 25°C. This procedure is illustrated in Problem 3.

PROBLEM 3.

The transistor 2N4425 is biased at the operating point V_{CE} = 5 V and I_C = 2 mA. Determine the hybrid small signal current gain parameter h_{fe} at the given operating point.

The transistor specification sheet provides a typical value for h_{fe}: h_{fe} = 180 at an operating point of V_{CE} = 10 V and I_C = 1 mA. Using the h parameter versus voltage correction curve, in the specification sheet, the normalized factor for h_{fe} at a V_{CE} of 5 V is 0.95. Next, referring to the h parameter versus I_C correction curve, the normalized factor is found for h_{fe} at I_C of 2 mA: h_{fe} = 1.1. The actual value of h_{fe} is:

$$h_{fe} = 0.95 \times 1.1 \times 180 = \mathbf{188}$$

The maximum collector power dissipation rating for a 2N217 transistor at 25°C is 150 mW. The derating factor in free air is 3.33 mW/°C increase in the ambient temperature above 25°C. Find the collector dissipation at 40°C, 55°C, and 100 °C. (R8-1)

POWER DISSIPATION CURVE

The power dissipation parameters P_T and P_C are of particular interest because they indicate how much power the transistor or its collector can dissipate without being destroyed. The majority of the power of the transistor is dissipated in the collector region. Therefore, it is not possible to distinguish between P_T and P_C and they can be considered to be equal.

The power dissipated in the collector region of a transistor connected in a common-emitter configuration is equal to the voltage across the transistor, V_{CE}, times the current

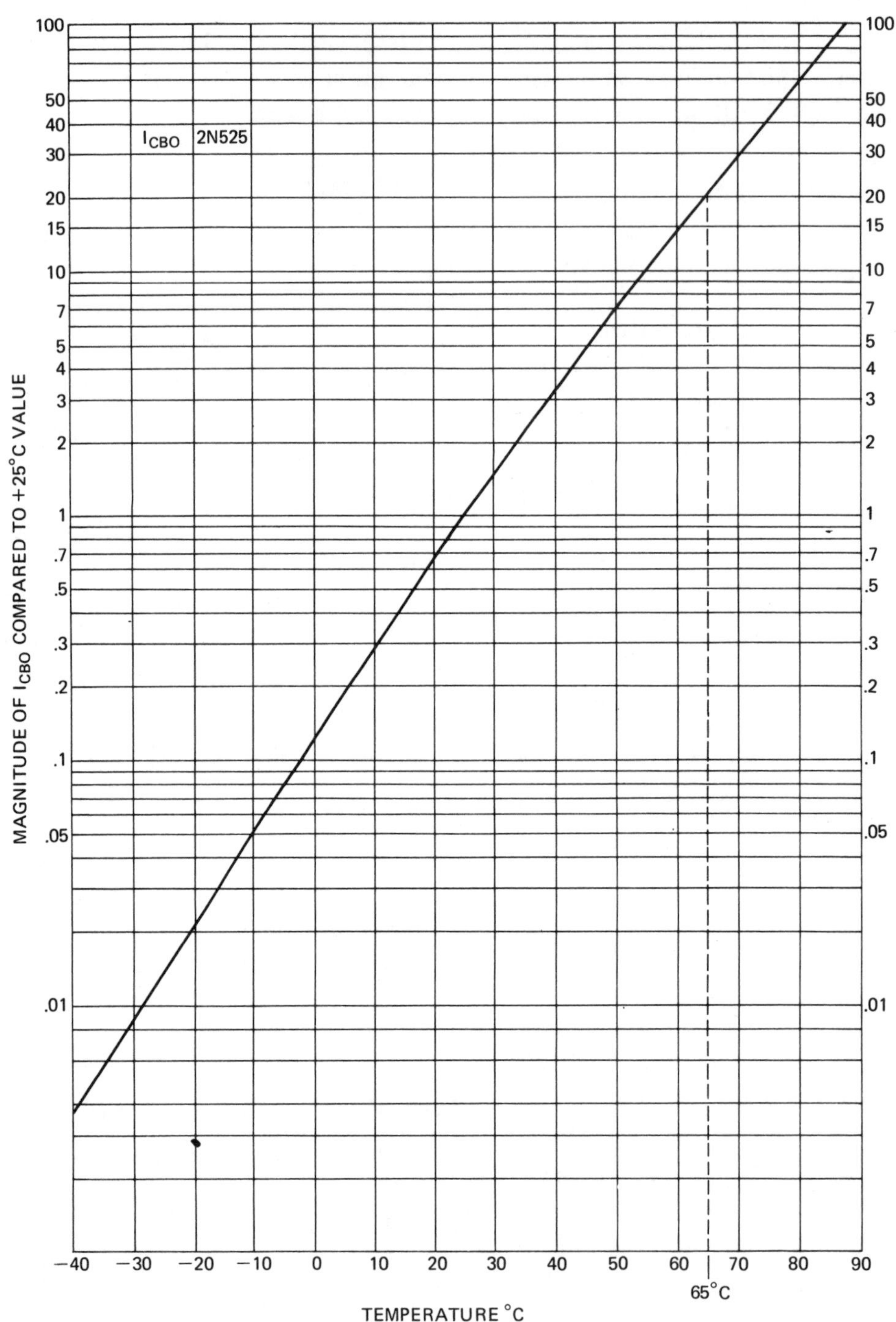

FIG. 8-3 COLLECTOR CUTOFF CURRENT VS. TEMPERATURE

through the collector, I_C. If the current through the collector is increased, the allowable voltage across the collector must decrease if P_C is to remain constant. This statement can be illustrated by plotting a power dissipation curve on the collector characteristics. If the power dissipation is a maximum for the transistor, then the curve becomes the *maximum power dissipation* curve. A power dissipation curve is plotted in Problem 4.

PROBLEM 4.

A 2N525 transistor is operated in free air at 25°C. Determine the maximum power dissipation. Draw the maximum power dissipation curve on the common-emitter collector characteristic curve.

Referring to the specification sheet for the 2N525 transistor, P_T = 225 mW.

$$P_T = P_C = V_{CE}\, I_C \qquad \text{Eq. 8.3}$$

where V_{CE} = voltage from the collector to the emitter

I_C = collector current

P_C = power dissipation of the collector, or the transistor.

To obtain the data for the plotting of the maximum power dissipation curve, let the collector-emitter voltage equal some convenient value. Calculate the collector current when P_C = 225 mW. A sample calculation is shown below.

Let V_{CE} = 5 V:

$$\text{then, } I_C = \frac{P_C}{V_{CE}} = \frac{225 \text{ mW}}{5 \text{ V}} = 45 \text{ mA}$$

The results of similar calculations are listed in the following table.

V_{CE} in volts	I_C in mA
10	22.50
15	15.00
20	11.25
25	9.00

The maximum power dissipation curve plotted from these values is shown in figure 8-4.

If the transistor operates in the region to the *left* of the maximum power curve in figure 8-4, then the power dissipation of the transistor is less than 225 mW. Similarly,

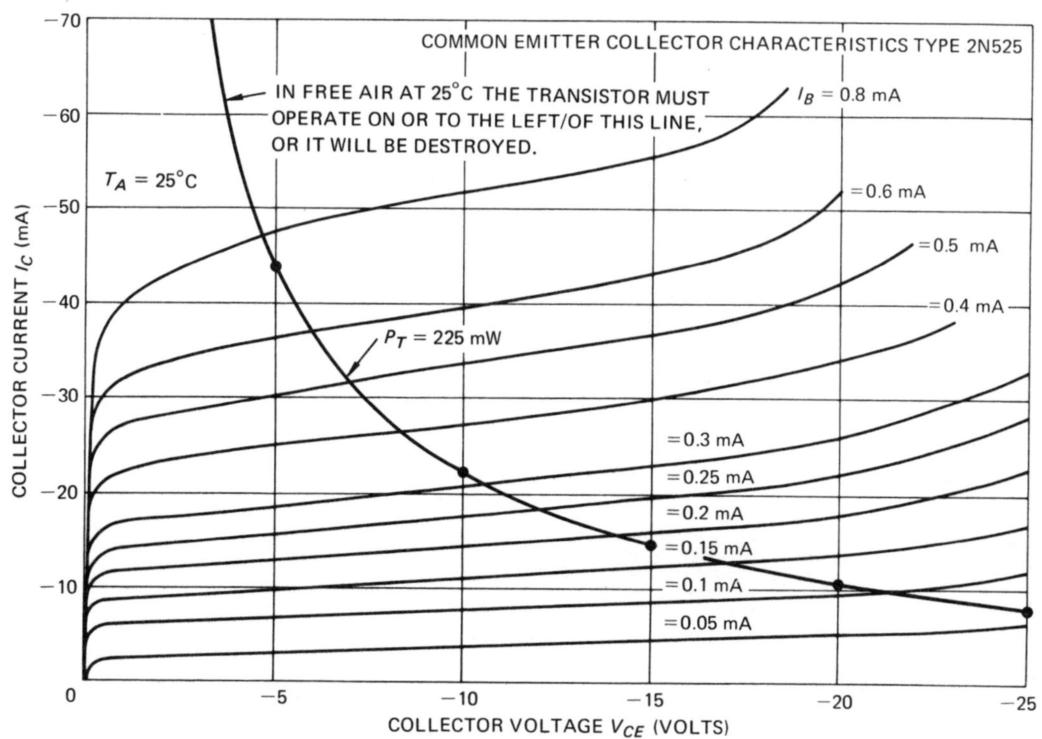

FIG. 8-4 MAXIMUM POWER DISSIPATION CURVE FOR P_T = 225 mW

operation of the transistor in the region to the *right* of the maximum power curve will exceed the power dissipation rating of 225 mW, causing the probable destruction of the transistor. Operation exactly on the maximum power curve is not recommended since an appreciable increase in temperature will cause the operation of the transistor to shift to the region to the right of the maximum power curve. As a result, the transistor will probably be destroyed. A safe rule of thumb is to operate the transistor at a value 30% *less* than the maximum power rating.

What purpose does the maximum power dissipation curve serve? (R8-2)

EXTENDED STUDY TOPICS

1. A transistor has a maximum collector power dissipation rating factor of 1W at 25°C. The derating factor in free air is 50 mW/°C increase in the ambient temperature above 25°C. Find the collector dissipation at 45°C, 60°C, and 150°.
2. Using the temperature curves for the 2N525 transistor, determine the leakage current I_{CBO} at -15°C, 0°C, 30°C, and 70°C.
3. Using the temperature curves for the 2N4424 transistor, determine every h parameter at -20°C, 0°C, 20°C, and 60°C.
4. Using the curves for the 2N4424 transistor, find the quantities h_{fe}, h_{re}, h_{rb}, and h_{ob} at the following values:

	V_{CE} in V	I_C in mA
a.	1	3
b.	10	5
c.	7	6
d.	3	10

5. Draw a 200-mW power dissipation curve for the 2N524 transistor on figure 8-1.
6. Draw a 300-mW power dissipation curve for the 2N4425 transistor on figure 8-2.

Unit 9 Transistor load lines

OBJECTIVES

After studying this unit, the student will be able to discuss and demonstrate an understanding of the basic principles of:

- Constructing a dc load line
- Establishing a quiescent operating point
- Constructing an ac load line

INTRODUCTION

The collector characteristics are very important when analyzing the operation of a transistor. The collector characteristics may be given by the manufacturer's specification sheets, or may be obtained by using a transistor curve tracer or by constructing the circuits given in units 5 and 7 and performing the necessary tests.

If a transistor is to be used as an amplifier, the transistor must be placed in a circuit with a load. The most widely used circuit for such an application is the common-emitter configuration shown in figure 9-1. We will limit our comments in this section to this configuration. If a particular operating point can be picked, V_{CE} and I_C or I_B, the collector characteristics will indicate what can be expected from the transistor amplifier. How is the operating point selected, and once it is selected, how are the necessary circuit components determined to maintain the operating point?

For most amplifier circuits, the transistor should operate in the active region of the characteristic curves. When an ac input signal is applied to the circuit, the output will vary around a center point. This point

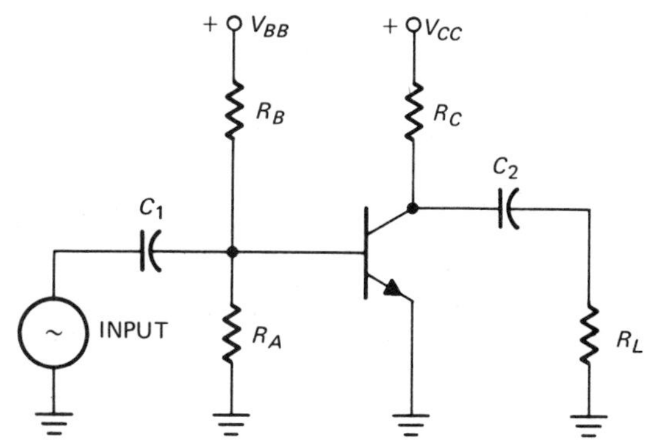

FIG. 9-1 TYPICAL COMMON-EMITTER AMPLIFIER CIRCUIT

is called the *operating point,* the *quiescent point,* or *Q point,* figure 9-2. To determine the operating point of a circuit, the ac input signal must be equal to zero. In other words, the operating point is determined by the dc voltages and currents that exist in the amplifier circuit. The operating point is usually specified by the collector to emitter voltage, V_{CE}, and the collector current, I_C. These symbols may be followed by the letter Q as shown in figure 9-5. A certain value of the base current, I_{BQ}, is required to obtain an operating point. Figure 9-1 shows the circuit parameters R_A, R_B, and V_{BB} which must be determined before the base current, I_{BQ}, can be found. The method of determining these parameters is known as biasing and will be analyzed in detail in the last units of this text.

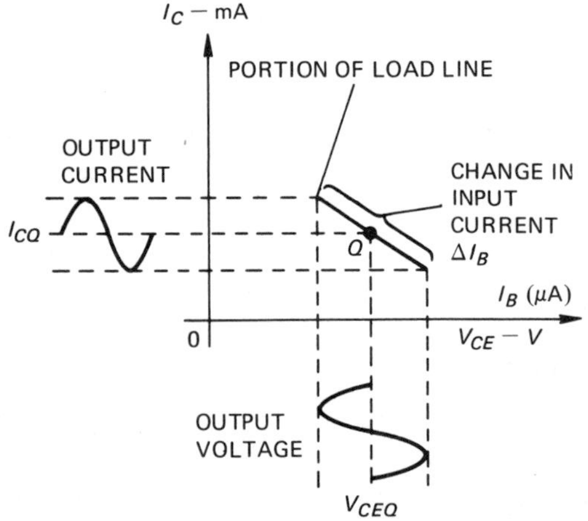

FIG. 9-2 AN OPERATING POINT WITH INPUT SIGNAL APPLIED

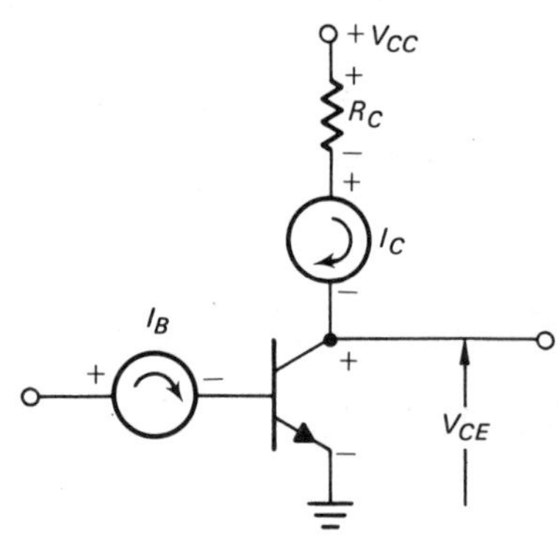

FIG. 9-3 DC EQUIVALENT CIRCUIT FOR FIG. 9-1

DC LOAD LINE

The operating point is determined from the dc circuit shown in figure 9-3. This circuit is the dc equivalent circuit of figure 9-1. (At this point, it is necessary to know only that a base current does exist, not the value of the base current.) The output of the circuit is taken from the collector to the emitter of the transistor. The transistor collector-emitter circuit is a closed loop, which includes the supply V_{CC}, the collector resistor R_C, the transistor collector-emitter voltage, and the return to ground. Eq. 9.1 is obtained by applying Kirchhoff's voltage law around this closed loop.

$$V_{CC} = I_C R_C + V_{CE} \qquad \text{Eq. 9.1}$$

From Eq. 9.1, and the collector characteristics, we can study the operation of a transistor amplifier circuit.

Eq. 9.1 is similar to equations of the form c = ax + by, such as 9 = 3x + y. To plot the equation 9 = 3x + y, first find the x and y intercepts. That is:

Let x = 0, then y = 9.

Let y = 0, then x = 3.

When the above points are plotted and then joined with a straight line, the graph shown in figure 9-4 results.

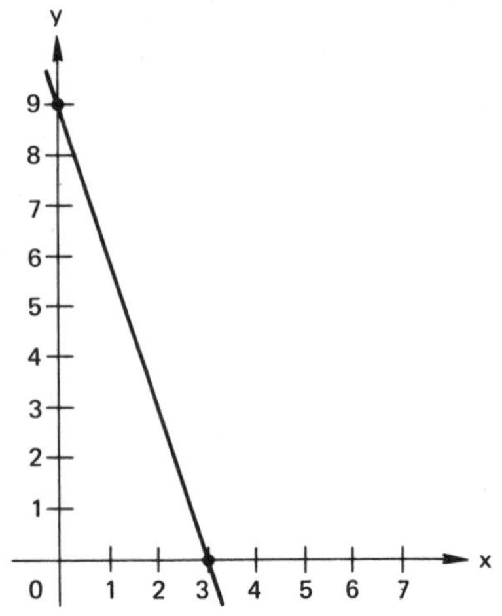

FIG. 9-4 PLOT OF EQUATION, 9 = 3x + y

This same procedure is followed to plot Eq. 9.1 on the output curves for a transistor.

Let $I_C = 0$, then $V_{CE} = V_{CC}$ (horizontal intercept).

Let $V_{CE} = 0$, then $I_C = \dfrac{V_{CC}}{R_C}$ (vertical intercept)

When these points are plotted on the output curves of a transistor and then joined by a straight line, the graph shown in figure 9-5 results.

This straight line graph is called the *dc load line.* The dc load represents the operating location for this transistor given the specific supply voltage, V_{CC}, and the collector resistor, R_C.

When determining the x and y intercepts, if V_{CC}/R_C should fall off the characteristic curve, the second point can be obtained by the following procedure. Select a voltage, V_1, far enough from V_{CC} so that an accurate straight line can be drawn. Divide V_1 by the dc load resistance, in this case R_C, to obtain a vertical distance $I_C = V_1/R_C$. Plot this point as shown on the curve in figure 9-5. Draw the dc load line through this new point. Note that a triangle is formed with a base distance equal to V_1 and a height of V_1/R_C. This triangle is similar to the triangle with a base of V_{CC} and a height of V_{CC}/R_C.

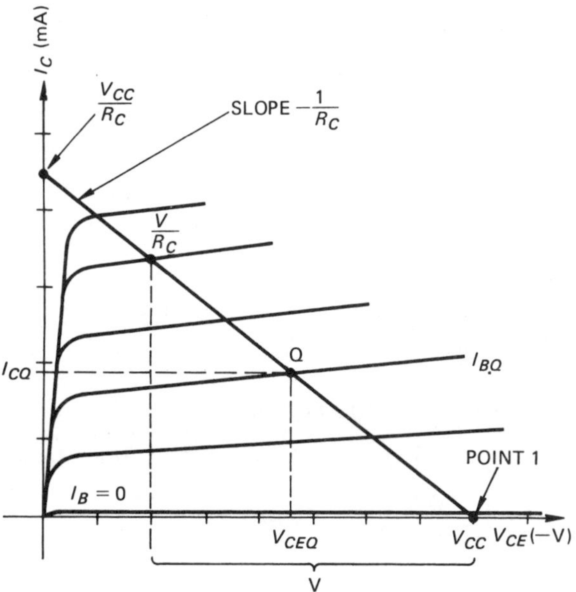

FIG. 9-5 DC LOAD LINE FOR THE DC LOAD R_C

The intersection of the dc load line with a dc base current, I_B, is called the operating point (Q), figure 9-5. The base current I_{BQ} is the current flowing in the base circuit when the ac signal is equal to zero. The selection of I_{BQ} is critical. If the base current is near zero, the transistor will operate near cutoff; if the base current is large, the transistor will operate near or in the saturation region. In figure 9-5, I_{BQ} is selected to be approximately in the center of the dc load line.

The following procedure summarizes how the dc load is plotted and how the operating point is located.

1. Point 1: Let $I_C = 0$, then $V_{CE} = V_{CC}$.
2. Determine the dc load, R_{dc}, and draw the dc equivalent circuit.
3. Point 2: Let $V_{CE} = 0$, then $I_C = \dfrac{V_{CC}}{R_{dc}}$

If I_C, as determined in step 3, falls off the characteristic curve, the following procedure is used to obtain a value for I_C.

Move from Point 1 to the left a distance equal to a convenient value of voltage V_1. Divide V_1 by the dc load to obtain the vertical distance

$$I_C = \frac{V_1}{R_{dc}}$$

4. Plot points 1 and 2 and join them with a straight line.
5. Locate the operating point at the intersection of the dc load line and the selected dc base current, I_{BQ}. The value

of I_{BQ} is selected to satisfy the transistor application.

The slope of the dc load is equal to $-1/R_C$. This can be seen more readily if the straight line equation is rearranged in the slope intercept form, y = mx + b. For this equation, m is the slope of the line and b is the y intercept. Eq. 9.1 expressed in the slope intercept form becomes:

$$I_C = \frac{-1}{R_C}(V_{CE}) + \frac{V_{CC}}{R_C} \qquad \text{Eq. 9.2}$$

Eq. 9.2 indicates that the slope of the line is $-1/R_C$. In addition, the intercept on the vertical or I_C axis is shown to be V_{CC}/R_C. An alternate method of plotting the dc load line is provided by Eq. 9.2. That is, start at the point $V_{CE} = V_{CC}$, and draw a line having a slope of $-1/R_C$.

PROBLEM 1.

The circuit shown in figure 9-6 has the following values for the parameters V_{CC}, R_C, and I_B:

	V_{CC} in volts	R_C in ohms	I_{BQ} in mA
a.	20	400	0.4
b.	15	500	0.2
c.	20	4000	0.03
d.	20	200	0.5

Using the collector characteristics for the 2N525 transistor, draw load lines for the given values of R_C and V_{CC}. Indicate on the graph the quiescent values for the collector voltage and current for the given values of I_{BQ}.

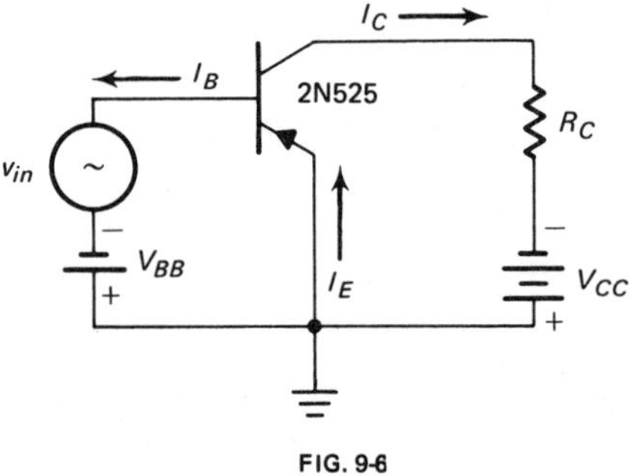

FIG. 9-6

When Kirchhoff's voltage law is applied around the collector circuit, $V_{CC} = I_C R_C + V_{CE}$.

a. 1. Point 1: Let $I_C = 0$, then V_{CE} = -20 V

2. R_{dc} = 400 ohms

3. Point 2: Let $V_{CE} = 0$, then I_C = $\frac{-20 \text{ V}}{400\ \Omega}$ = -50 mA

4. Plot points 1 and 2 and join them with a straight line as shown in figure 9-7.

5. The operating point is the intersection of I_{BQ} = 0.4 mA and the dc load line. At this point, V_{CEQ} = -9V, and I_{CQ} = -27 mA.

b. 1. Point 1: Let $I_C = 0$, then V_{CE} = -15 V

2. R_{dc} = 500 ohms

3. Point 2: Let $V_{CE} = 0$, then I_C = $\frac{-15 \text{ V}}{500\ \Omega}$ = -30 mA

4. Plot points 1 and 2 and join them with a straight line as shown in figure 9-7.

5. The operating point is the intersection of I_{BQ} = 0.2 mA and the dc load line. At this point, V_{CEQ} = -8 V, and I_{CQ} = -14 mA.

c. 1. Point 1: Let $I_C = 0$, then V_{CE} = -20 V

2. R_{dc} = 4000 ohms

3. Point 2: Let V_{CE} = 0 V, then I_C = $\frac{-20 \text{ V}}{4000\ \Omega}$ = -5 mA

4. Plot points 1 and 2 and join them with a straight line as shown in figure 9-8.

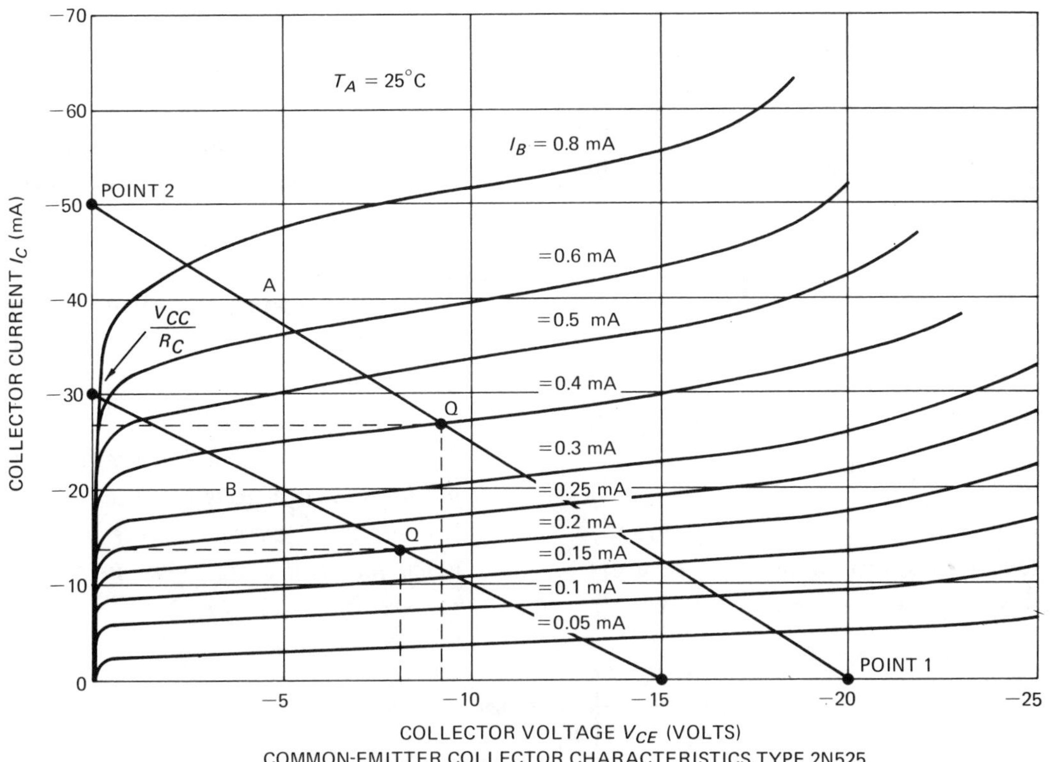

FIG. 9-7 SOLUTION FOR PARTS A AND B OF PROBLEM 1

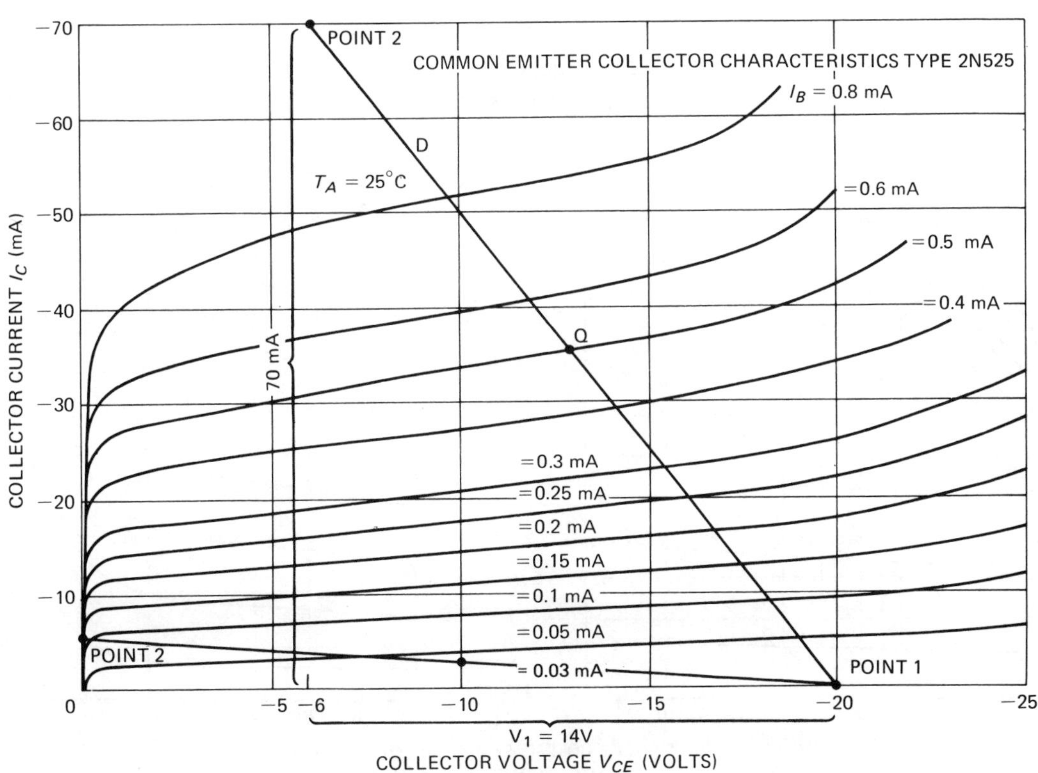

FIG. 9-8 SOLUTION FOR PARTS C AND D OF PROBLEM 1

5. The load line is low on the characteristic curve; therefore, the Q point values must be estimated as accurately as possible. The operating point is the intersection of I_{BQ} = 0.03 mA and the dc load line. At this point, V_{CEQ} = -10 V, and I_{CQ} = -2.5 mA.

d. 1. Point 1: Let I_C = 0, then V_{CE} = -20 V

2. R_{dc} = 200 ohms

3. Point 2: Let V_{CE} = 0, then I_C = $\frac{-20\text{ V}}{200\ \Omega}$ = -100 mA

Since 100 mA is not on the graph, move to the left of point 1 an arbitrary distance V_1 equal to 14 V. At this point, V_{CE} = -6 V. Then,

$$I_C = \frac{-14\text{ V}}{200\ \Omega} = -70\text{ mA}$$

4. Plot points 1 and 2 and join them with a straight line as shown in figure 9-8.

5. The operating point is the intersection of I_{BQ} = 0.5 mA and the dc load line. At this point, V_{CEQ} = -13 V, and I_{CQ} = -35 mA.

The point on a dc load line that is plotted on the voltage axis of the transistor characteristic curve is determined by which physical component in an amplifier circuit? (R9-1)

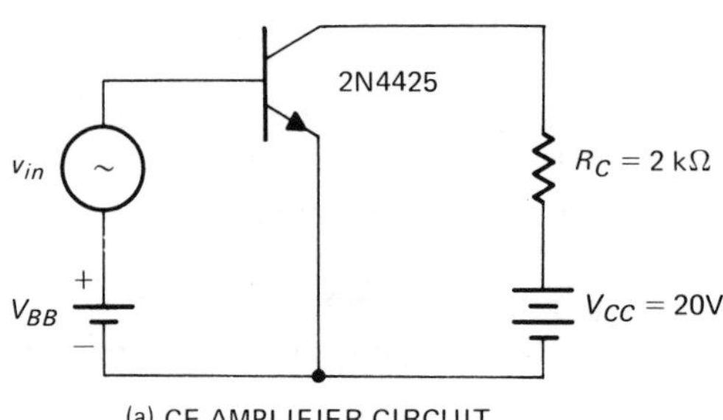

(a) CE AMPLIFIER CIRCUIT

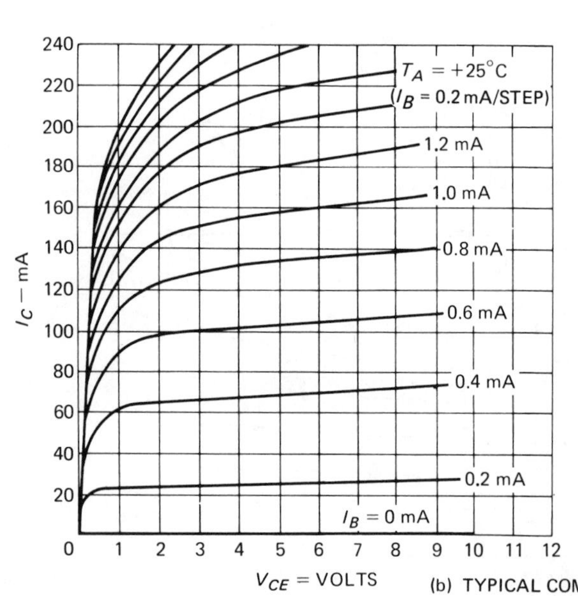

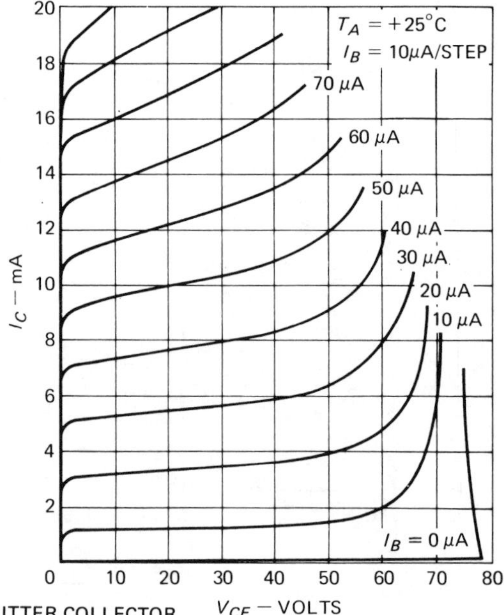

(b) TYPICAL COMMON-EMITTER COLLECTOR CHARACTERISTICS 2N4425

FIG. 9-9

Write the equation that determines the point on a dc load line that is plotted on the current axis of the transistor characteristic curve. (R9-2)

Name the three parameters that determine the Q point in figure 9-5. (R9-3)

On figure 9-9b, draw the dc load line for the circuit in figure 9-9a. (R9-4)

AC LOAD LINE

Figure 9-10 illustrates the two paths that the ac collector current can take in the common emitter circuit of figure 9-1. One current path is through the capacitor, which acts as a *short* circuit to ac, and through resistor R_L to ground. The other current path is through resistor R_C and then through the dc power supply, which acts like a short circuit to ac, to ground. The power supply filter circuit, which is shown within the dotted outline in figure 9-10, usually contains a capacitor. It is apparent in the circuit in figure 9-10 that the ac load on the transistor is the equivalent resistance of R_C in parallel with R_L.

Is there any common point between the ac and the dc load lines? When the ac signal is equal to zero, the only load on the transistor is the dc load. The collector-emitter voltage and the collector current can be found by reading the values of these quantities at the Q point or operating point on the collector characteristics. Therefore, one point of the ac load line is the Q point. The second point of the ac load line is determined using a procedure which is similar to that for finding the dc load line. For this case, however, the *ac load* on the transistor must be used. The procedure for plotting the ac load line is given below.

1. Point 1 is the Q point.
2. Calculate the ac load in the collector circuit. From figure 9-10:

$$R_{ac} = R_C \,||\, R_L = \frac{R_C \; R_L}{R_C + R_L}$$

3. To obtain a second point for the ac load line, select some convenient value of the voltage V_{CEQ} to move from the Q point to the vertical axis, figure 9-11, page 156.
4. Divide the value of the voltage V_{CEQ} in step 3 by the ac load found in step 2.

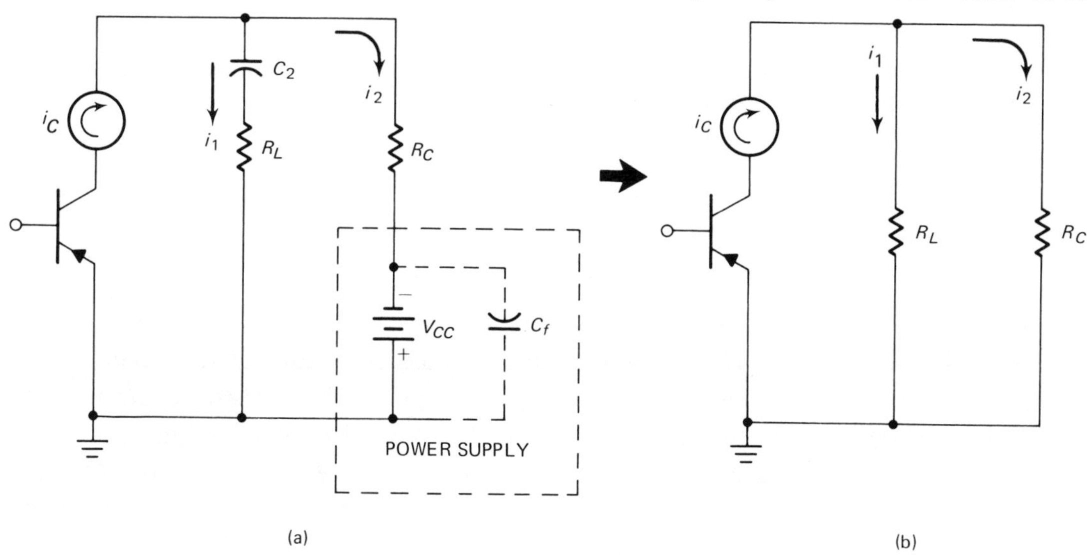

FIG. 9-10 AC COLLECTOR CURRENT PATHS AND AC EQUIVALENT CIRCUITS

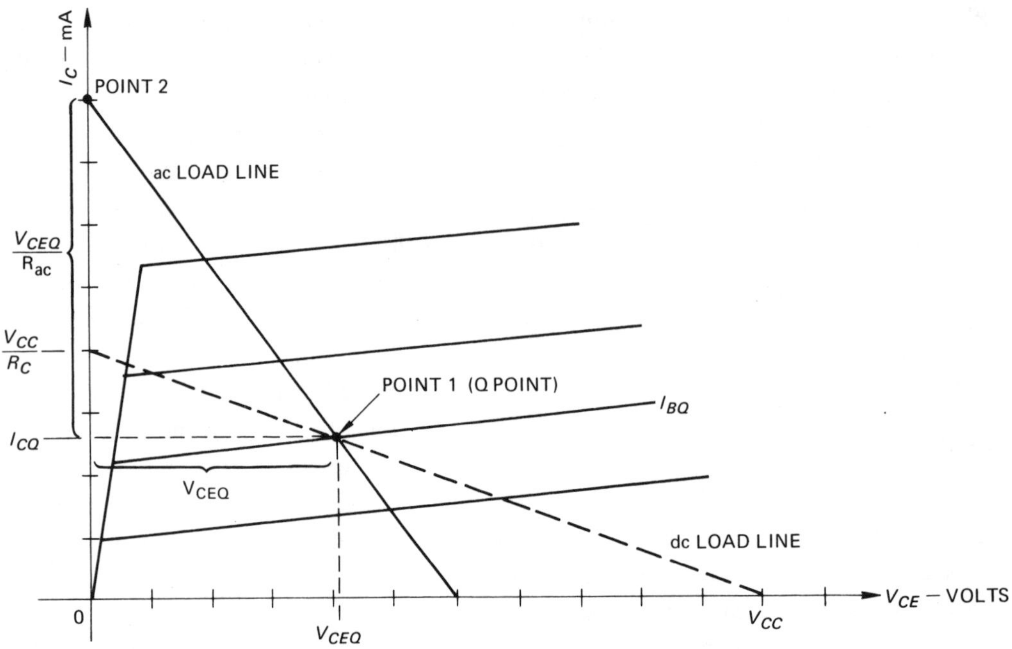

FIG. 9-11 CONSTRUCTION OF THE AC LOAD LINE

The resulting current is V_{CEQ}/R_{ac}. This current value is added to I_{CQ} and the total current value is plotted as point 2 of the ac load line, figure 9-11. If the second point is not on the graph, select a smaller value of voltage when moving to the left of the Q point. Refer to the summary of the procedure for plotting dc load lines.

The slope method also can be used to plot the ac load line. The slope of the ac load is $-1/R_{ac}$. Start at the Q point and draw a line with a slope equal to $-1/R_{ac}$ through the Q point.

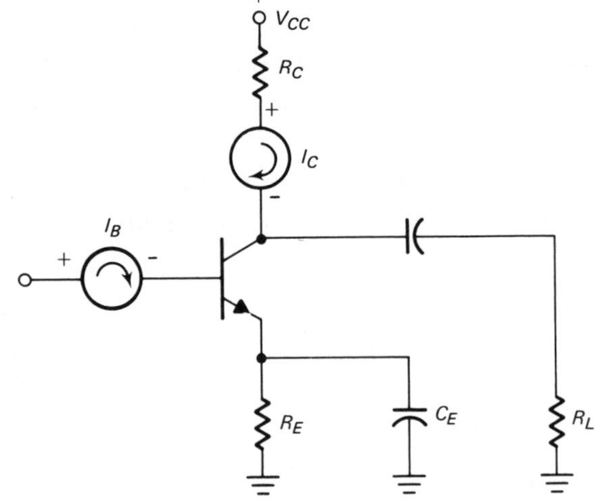

FIG. 9-12 COMMON-EMITTER CIRCUIT WITH EMITTER RESISTOR

An emitter resistor may be added to the common-emitter amplifier circuit to provide stability, figure 9-12. To prevent this resistor from affecting the ac circuit, it is bypassed with a capacitor, C_E. In other words, the capacitor C_E in parallel with the emitter resistor appears as a short circuit at the operating ac frequency. The previous method for drawing the load lines also applies to the circuit in figure 9-12. However, now the dc load in the collector-emitter circuit is $R_C + R_E$. Both of these resistors must be considered when the dc load line is drawn, as shown in figure 9-13a. The emitter resistor is not part of the ac load since the resistor is bypassed by a capacitor. The ac load line is shown in figure 9-13b.

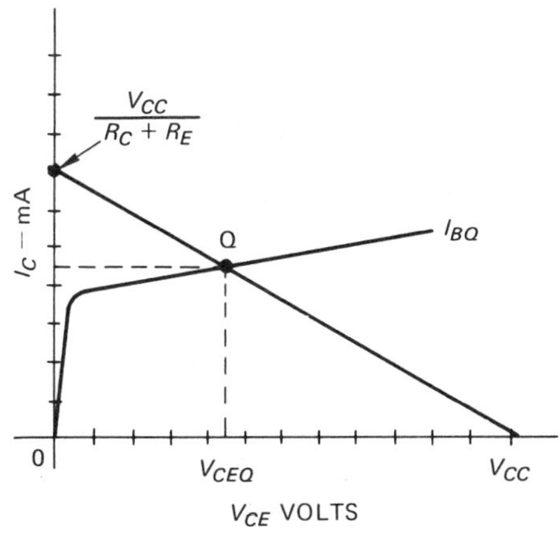

FIG. 9-13a DC LOAD LINE FOR CIRCUIT WITH EMITTER RESISTOR

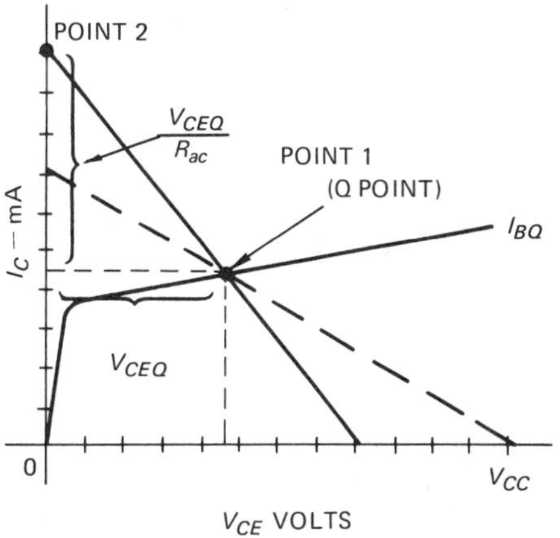

FIG. 9-13b AC LOAD LINE FOR CIRCUIT WITH EMITTER RESISTOR

PROBLEM 2.

For the circuit shown in figure 9-14, assume that I_B = 0.13 mA when no ac signal is applied.

a. Find the operating point.

b. Draw the ac load line.

c. Find the peak ac signal value of I_B when the transistor is in the cutoff region and when it is in the saturation region.

a. To find the operating point, first draw the dc load line.

1. Point 1: Let $I_C = 0$, then $V_{CE} = -20$ V.
2. $R_{dc} = R_C = 1\ k\Omega$
3. Point 2: Let $V_{CE} = 0$, then $I_C = \dfrac{-20\ V}{1\ k\Omega} = -20$ mA
4. Plot points 1 and 2 and join them with a straight line as shown in figure 9-15.

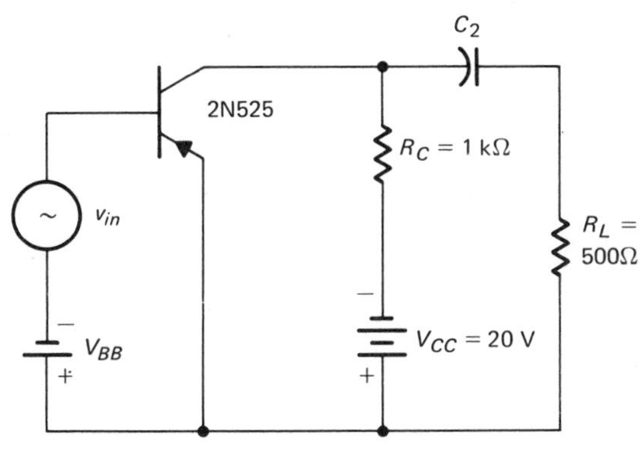

FIG. 9-14

5. The operating point is the intersection of I_{BQ} = 0.13 mA and the dc load line. At this point, V_{CEQ} = -10 V, and I_{CQ} = -10 mA

b. Draw the ac load line.

1. Point 1: The Q point is common to both the dc and ac load lines. The Q point is point 1 on the ac load line.
2. $R_{ac} = R_C \,||\, R_L = \dfrac{1000 \times 500}{1000 + 500} = 333\,\Omega$
3. Point 2: Move to the left of the Q point by an amount equal to V_{CEQ} volts. This move places point 2 on the I_C axis. $I_C = \dfrac{-10\ V}{333\ \Omega} = -30$ mA up from I_{CQ}.
4. Plot points 1 and 2 and join them with a straight line as shown in figure 9-15.

c. The Q point is located at I_{BQ} = 0.13 mA. This is the value of the peak ac signal required to drive the transistor into cutoff. The characteristic curve shows that I_B = 0.8 mA is the maximum base current that can be applied to the transistor. The peak ac signal value required to drive the transistor into saturation is 0.80 mA -0.13 mA = 0.67 mA.

Name one point that the ac and dc load lines always have in common. (R9-5)

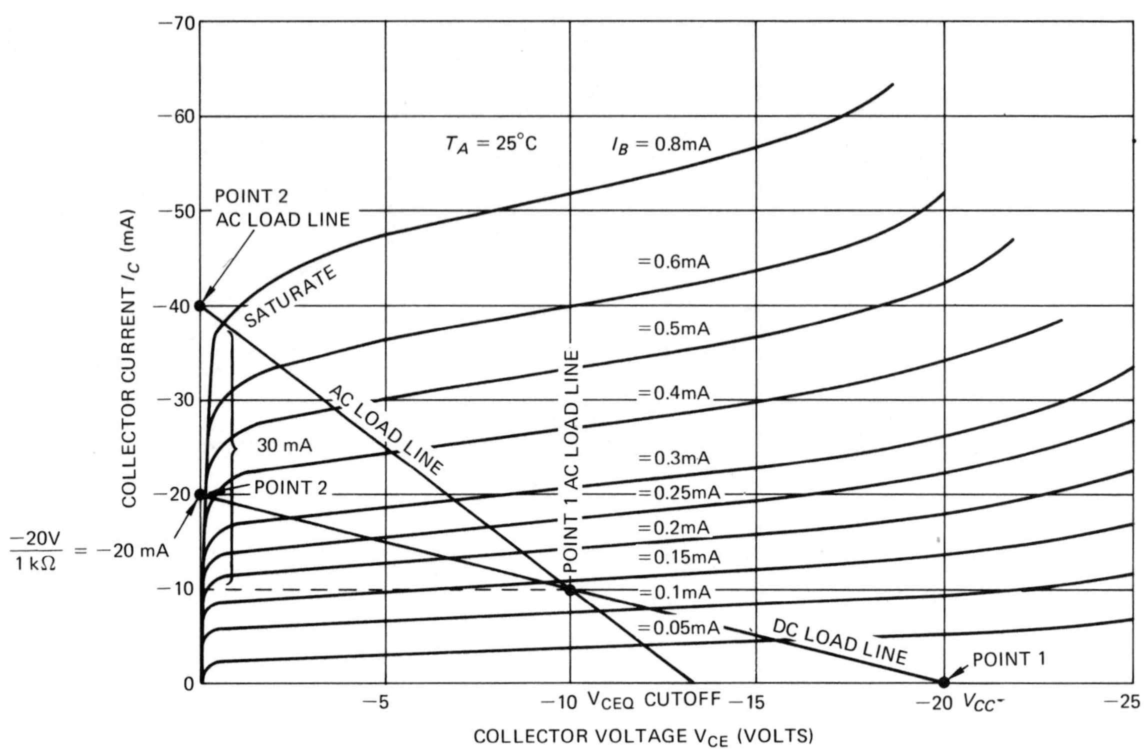

FIG. 9-15

LABORATORY EXCERCISE 9-1: LOAD LINE EXPERIMENT

PURPOSE

- To construct a dc load line and then compare theoretical and laboratory test results.
- To construct ac load lines.
- To observe the effect of a change in load resistors on the voltage gain of an amplifier circuit.

MATERIALS

1 NPN transistor, 2N4425, 2N3405 or equivalent
1 Dc power supply, 12 V
1 VTVM, or solid-state voltmeter
1 Ac VTVM (use an oscilloscope if an ac VTVM is not available)
2 VOMs, 20,000 ohms per volt
1 Signal generator
1 Oscilloscope
2 Electrolytic capacitors, 25μF, 50 V
1 Resistor, 300 Ω, 1/2W
1 Resistor, 150 Ω, 1/2 W
1 Resistor, 100 Ω, 1/2 W

1 Resistor, 10 k Ω, 1/2 W
1 Linear Potentiometer, 150 k Ω, 2 W
1 Switch, SPST, to connect the power supply to the circuit

PROCEDURE

A. 1. On figure 9-16, draw the collector power dissipation curve for 360 mW. If the CE collector characteristics were drawn for this transistor in Unit 7, use the previous curves instead of figure 9-16.

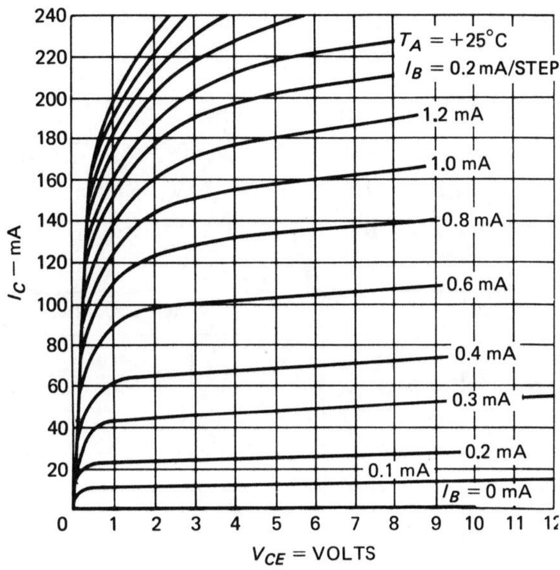

FIG. 9-16 CE COLLECTOR CHARACTERISTICS 2N4425

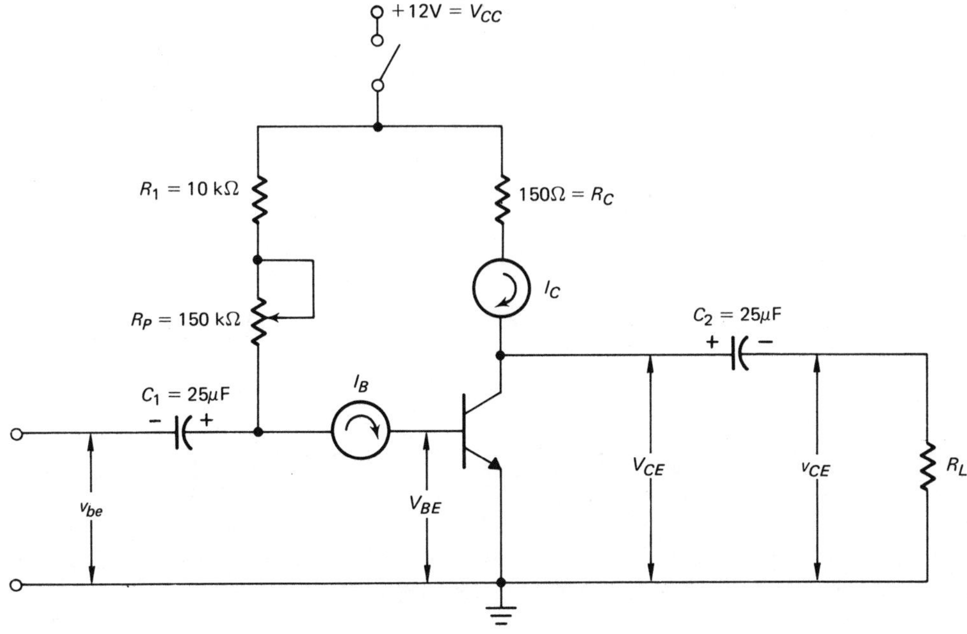

FIG. 9-17 EXPERIMENTAL CE AMPLIFIER CIRCUIT

2. Calculate the intercepts for the dc load line for the circuit shown in figure 9-17.

   $I_C = 0, V_{CE} = ?$

   $V_{CE} = 0, I_C = ?$

   Draw this dc load line on the CE collector characteristics used in step A.

3. Assume that a silicon transistor is used in the circuit. Calculate the minimum and maximum base current for figure 9-17.

   $I_{B\,(min)} = ?$ $I_{B(max)} = ?$

4. From the dc load line drawn in step B.2., determine the values of I_C and V_{CE} when I_B equals 0.1 mA, 0.2 mA, 0.3 mA, and 0.4 mA. Determine I_B and I_C for $V_{CE} = 6$ V. Record these calculations in the theoretical results section of Table 9-1.

TABLE 9-1

Theoretical Results			Actual Laboratory Results			
I_B in mA	I_C in mA	V_{CE} in V	I_B in mA	I_C in mA	V_{CE} in V	V_{BE} in V
0.1			0.1			
0.2			0.2			
0.3			0.3			
0.4			0.4			
		6			6	

5. Assume the Q point is at the intersection of the dc load line and $V_{CE} = 6$ V. If $R_L = 300\ \Omega$ in figure 9-17, determine the ac load (R_{ac}) of the transistor and the two points necessary for the construction of the ac load line. Draw the ac load line on figure 9-16.

   $R_{ac} =$

   Point 1:

   Point 2:

6. Repeat step 5 for $R_L = 100\ \Omega$. Find R_{ac} and points 1 and 2.

   $R_{ac} =$

   Point 1:

   Point 2:

B. Connect the circuit as shown in figure 9-17 but omit R_L. (R_L will be added in a later step.) The power supply switch should be open and the potentiometer should be set for the maximum resistance.

C. 1. Close the power supply switch and adjust the potentiometer for $I_B = 0.1$ mA.

2. Measure I_C, V_{CE}, and V_{BE} and record these values in Table 9-1 in the actual laboratory results section.

3. Repeat steps C.1 and C.2 for I_B = 0.2 mA, 0.3 mA, and 0.4 mA.

D. Adjust the potentiometer so that V_{CE} = 6 V. Measure I_B, I_C, and V_{BE} and record the values in Table 9-1 in the actual laboratory results section.

E. 1. Using the measured values recorded in Table 9-1, determine the change in the collector current, the collector to emitter voltage, and the base-emitter voltage for a base current change from 0.2 mA to 0.3 mA.

2. Calculate the current gain and the voltage gain. Record the results of the calculations in this step in Table 9-2.

TABLE 9-2

ΔI_B	ΔI_C	ΔV_{CE}	ΔV_{BE}	A_i	A_v
0.3 mA - 0.2 mA = 0.1 mA	mA	V	V		

F. 1. Connect a 300-Ω load resistor as shown in figure 9-17.

2. Adjust the potentiometer for V_{CE} = 6 V.

3. Connect an oscilloscope across the load resistor to observe v_{ce}.

4. Apply 1-kHz input sine wave from the signal generator, and increase the input signal until the output sine wave distorts. Then decrease the input signal until a nondistorted sine wave appears across the output. The input signal will be small, therefore a voltage divider may be required across the signal generator.

5. Using an ac VTVM, measure the input voltage, v_{be} and the output voltage, v_{ce}, and record these values in Table 9-3. If an ac VTVM is not available, use an oscilloscope and record the peak-to-peak readings in Table 9-3.

TABLE 9-3

R_L in Ω	v_{be} in V	v_{ce} in V	A_v
300			
100			

6. Calculate the voltage gain and record this value in Table 9-3.

G. Repeat step F for a 100-Ω resistor.

DISCUSSION QUESTIONS

1. Why is the collector power dissipation curve drawn before the load lines are constructed?

2. Why is a dc load line drawn?

3. How do the theoretical and actual results of Table 9-1 agree? Discuss any differences.

4. Why is an ac load line drawn?

5. How do the voltage gains found in steps F and G compare? Are they different? Should they be different? Refer to the ac load lines drawn in step A of the Procedure.

6. How do the voltage gains found in steps F and G compare with the voltage gain found in step E? Discuss reasons for any differences.

7. Using the load lines drawn in step A of the Procedure, determine the current gain of the amplifier. How does this gain compare with the gain determined in step E?

8. Can the voltage gain be determined from the load lines? Discuss the answer.

EXTENDED STUDY TOPICS

1. On figure 9-9b, draw the dc load line for the circuit in figure 9-9a for the following values:

   a. V_{CC} = 40 V and R_C = 8 k Ω

   b. V_{CC} = 10 V and R_C = 10 k Ω

2. For the circuit in figure 9-18a:

   a. Draw the dc load line on figure 9-18b.

   b. Locate the Q point.

   c. Draw the ac load line on figure 9-18b.

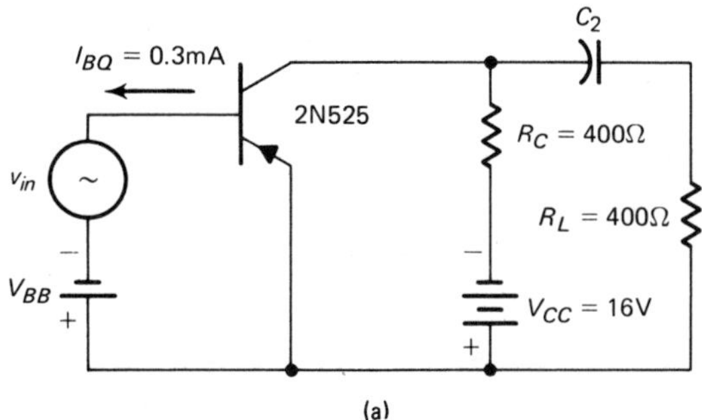

(a)

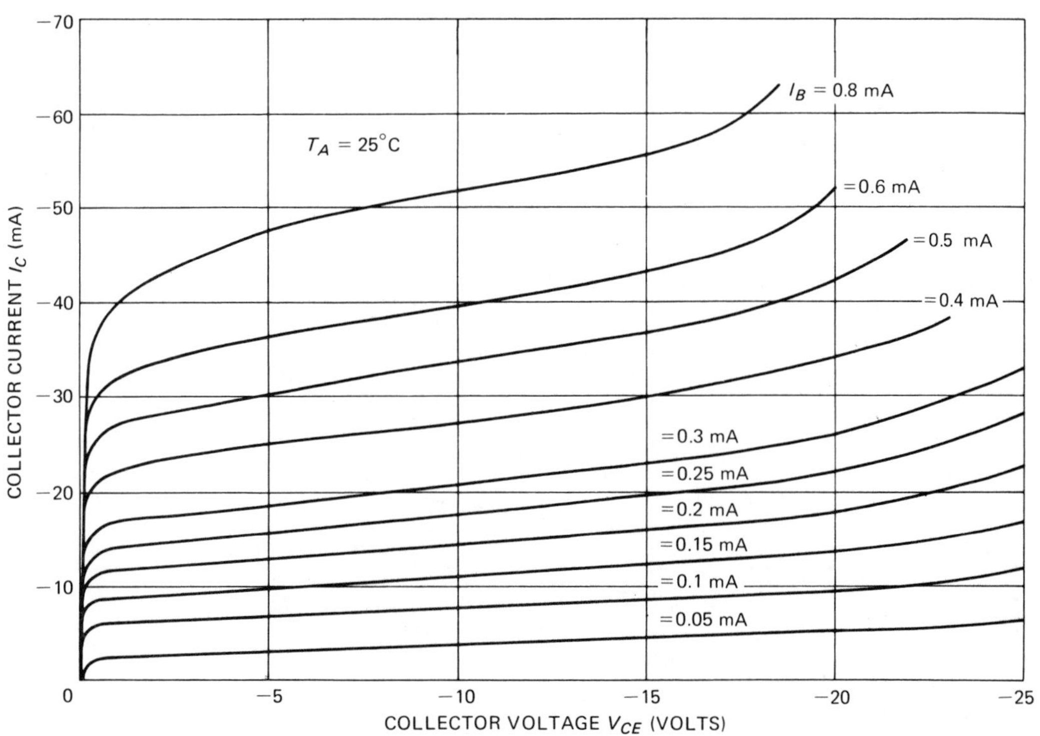

(b) CE CHARACTERISTIC CURVE FOR 2N525

FIG. 9-18

3. For the circuit in figure 9-19a:

   a. Draw the dc load in figure 9-19b.

   b. Locate the Q point.

   c. Draw the ac load line on figure 9-19b.

   d. At what peak ac base current will the transistor cut off?

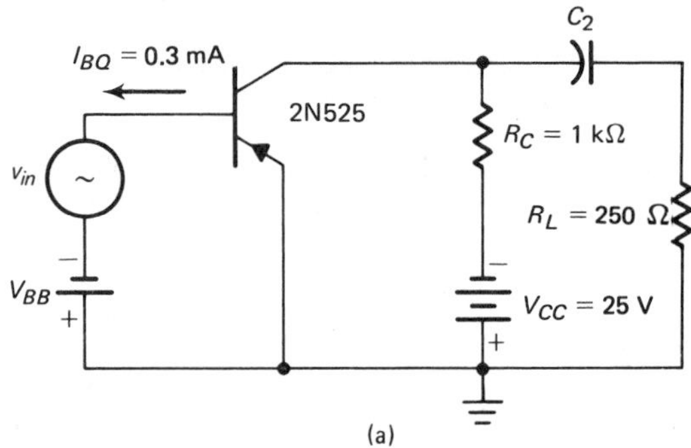

(a)

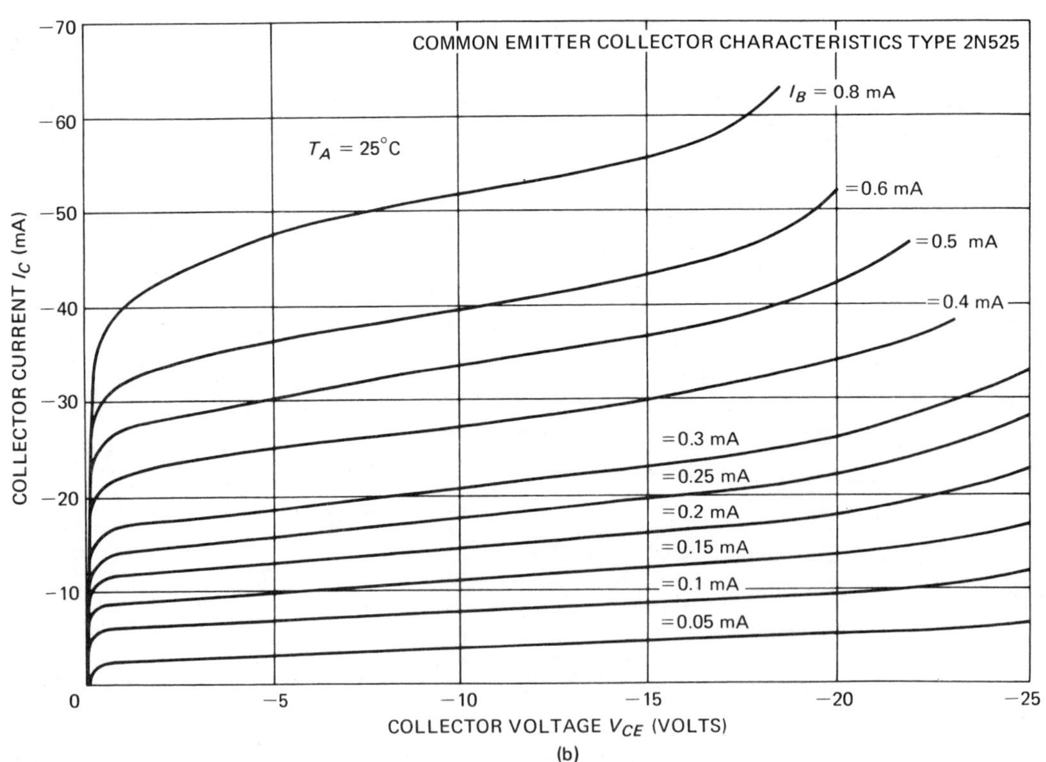

(b)

FIG. 9-19

4. For the circuit in figure 9-20a.

   a. Draw the dc load line on figure 9-20b.

   b. Locate the Q point.

   c. Draw the ac load line on figure 9-20b.

   d. At what peak ac base current will the transistor cut off?

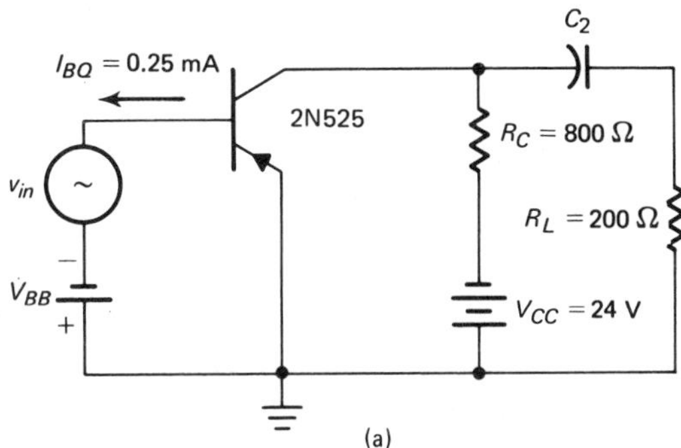

(a)

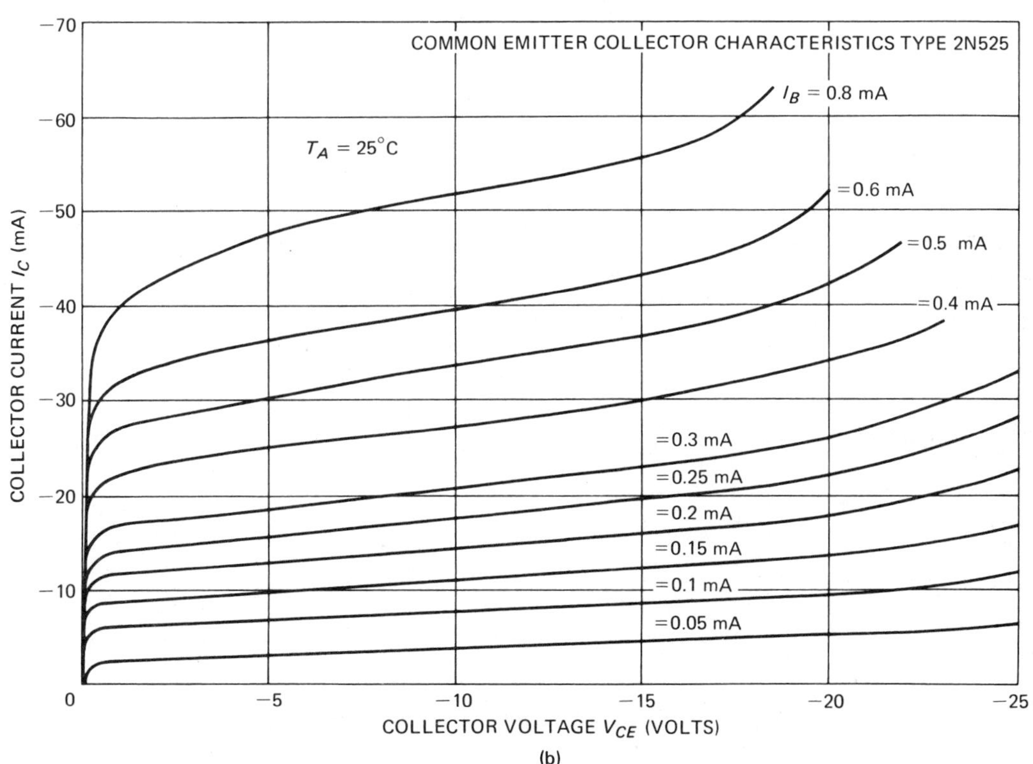

(b)

FIG. 9-20

5. For the circuit in figure 9-21a:

   a. Draw the dc load line on figure 21b.

   b. Locate the Q point.

   c. Draw the ac load line on figure 9-21b.

   d. At what peak ac base current will the transistor cut off?

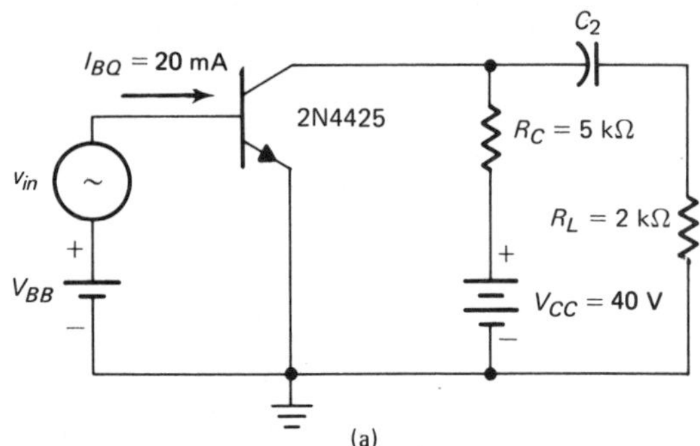

(a)

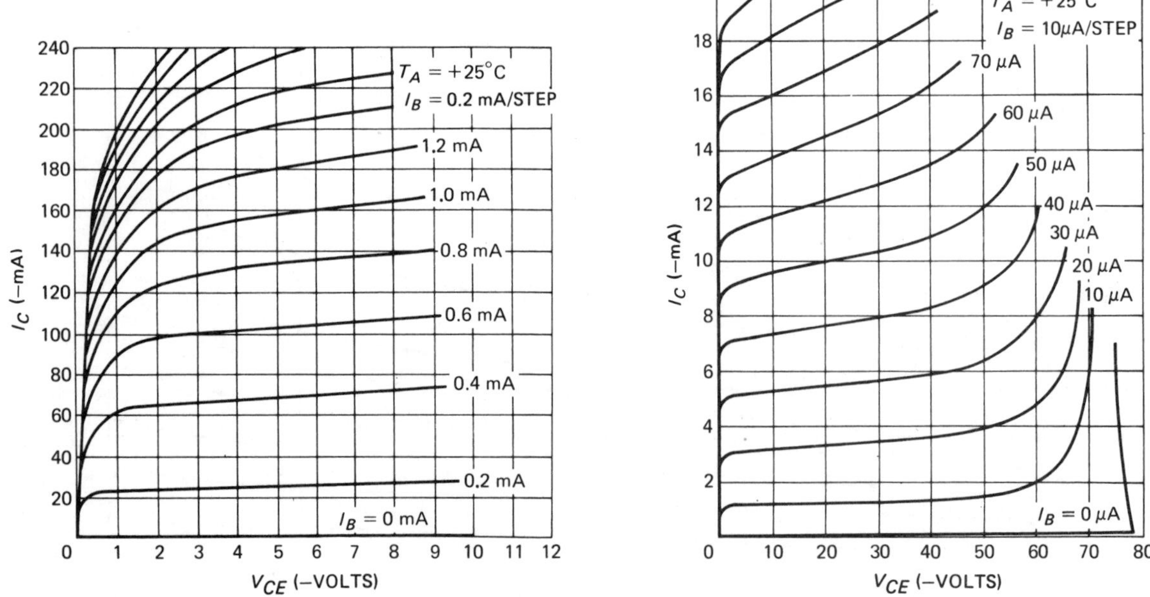

(b) TYPICAL COMMON EMITTER COLLECTOR CHARACTERISTIC 2N4425

FIG. 9-21

6. For the circuit in figure 9-22a:

   a. Draw the dc load line on figure 9-22b.

   b. Locate the Q point.

   c. Draw the ac load line on figure 9-22b.

   d. At what peak ac base current will the transistor cut off?

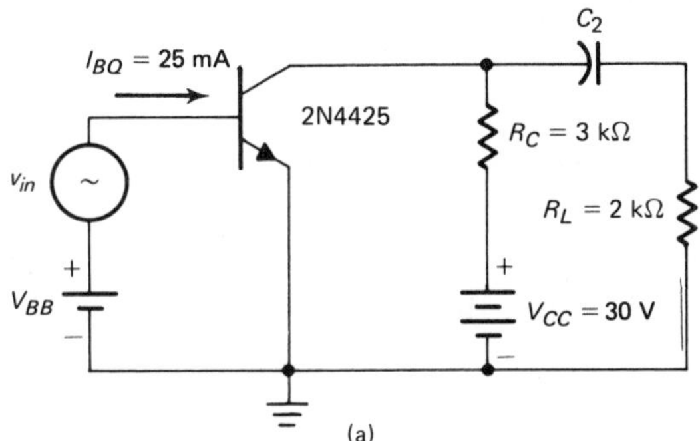

(a)

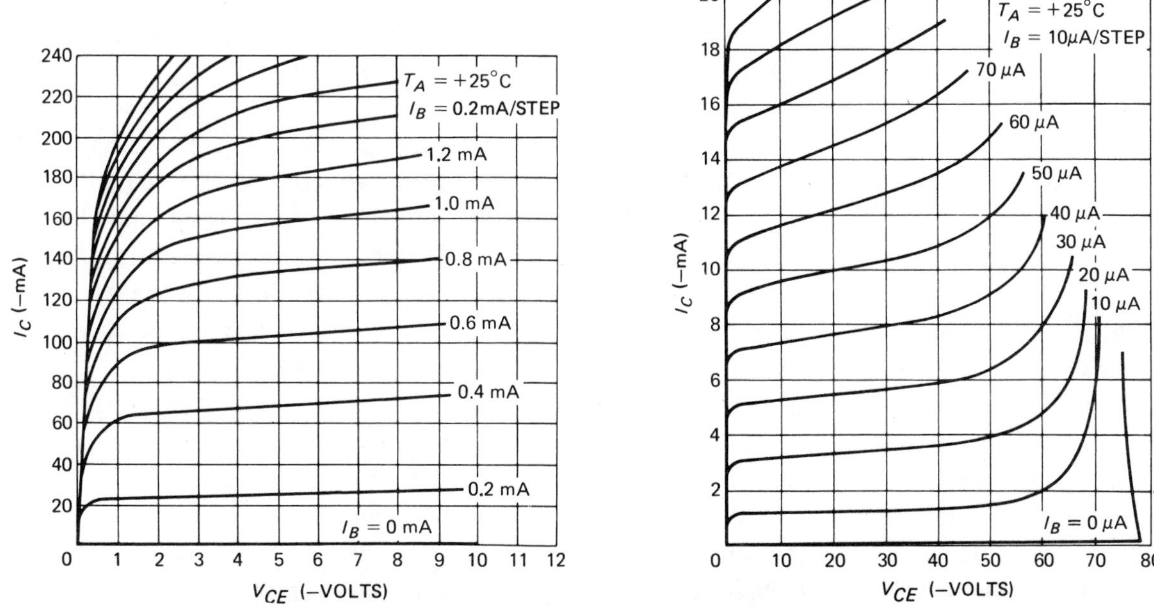

(b) TYPICAL COMMON-EMITTER COLLECTOR CHARACTERISTIC 2N4425

FIG. 9-22

Unit 10

Transistor biasing for common-emitter circuits

OBJECTIVES

After studying this unit, the student will be able to discuss and demonstrate an understanding of the basic principles of:

- Linear and nonlinear amplifiers
- The three limitations in locating the Q point in the active region
- Constructing and analyzing the following common-emitter bias circuits: fixed biased, and emitter-bias with single base resistor

INTRODUCTION

The dc operating conditions of an amplifier circuit are the dc voltages and currents that exist at the various circuit locations when there is no ac input signal. The biasing of a circuit is the selection of the proper components and dc supplies to establish these dc operating conditions so that the proper circuit operation is obtained. When a transistor is used as an amplifier, the circuit is usually designed so that the transistor will operate in the active region. For operation in the active region, the CE amplifier base circuit parameters control the collector circuit parameters.

The principle of amplification is one of the most important in electronics. However, no matter how well a circuit is designed to amplify an input signal, if the circuit does not have the correct bias, the circuit will not amplify properly. The circuit may amplify, but the output signal may be distorted. In other words, the output signal is *not* a replica of the input signal. This type of transistor amplifier is called a *nonlinear* amplifier circuit.

In figure 10-1, a simple ac signal in the form of base current i_b is applied to a transistor. The base current sine wave varies along the load line from the Q point at 20 μA to a maximum of 30 μA and a minimum of 10 μA. When the base current is 30 μA, the collector current is 5 mA. When the base current is 10 μA, the collector current is 1 mA. A base current change of 20 μA produces a collector current change of 4 mA. As a result, the current gain is equal to the change in the collector current divided by the change in the

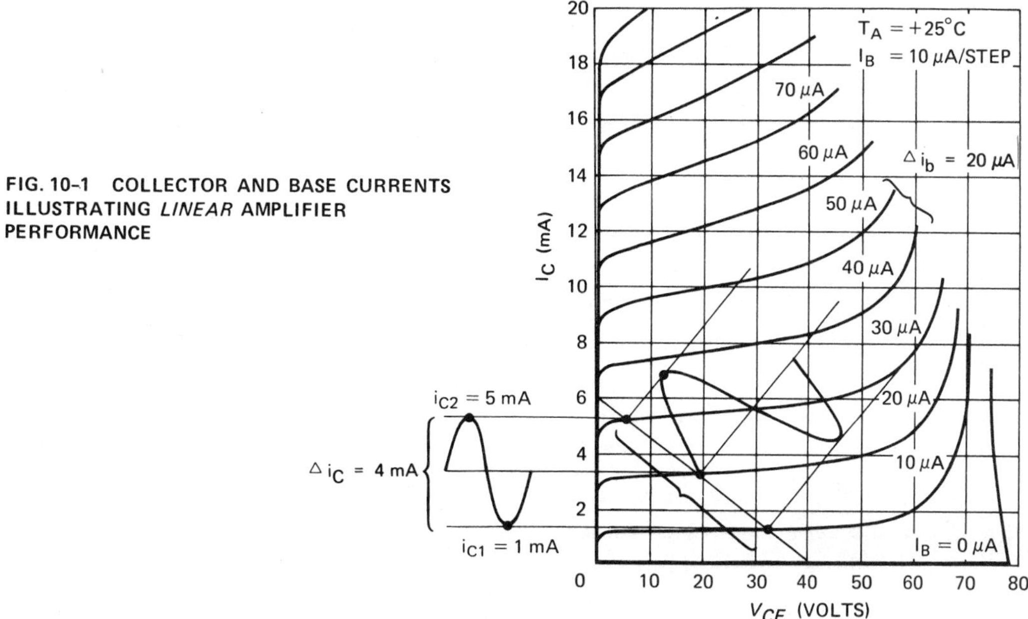

FIG. 10-1 COLLECTOR AND BASE CURRENTS ILLUSTRATING *LINEAR* AMPLIFIER PERFORMANCE

base current, or 4000 μA/20 μA = 200. These calculations can be made because the transistor amplifier is linear; that is, the output current i_c and the output current i_b are both sine waves.

Not all amplifier circuits, however, are linear. For nonlinear amplifier circuits, when the base current is a sine wave, the collector current is not a sine wave. To understand how this condition can exist, assume that the Q point in figure 10-1 is moved to a new location as shown in figure 10-2.

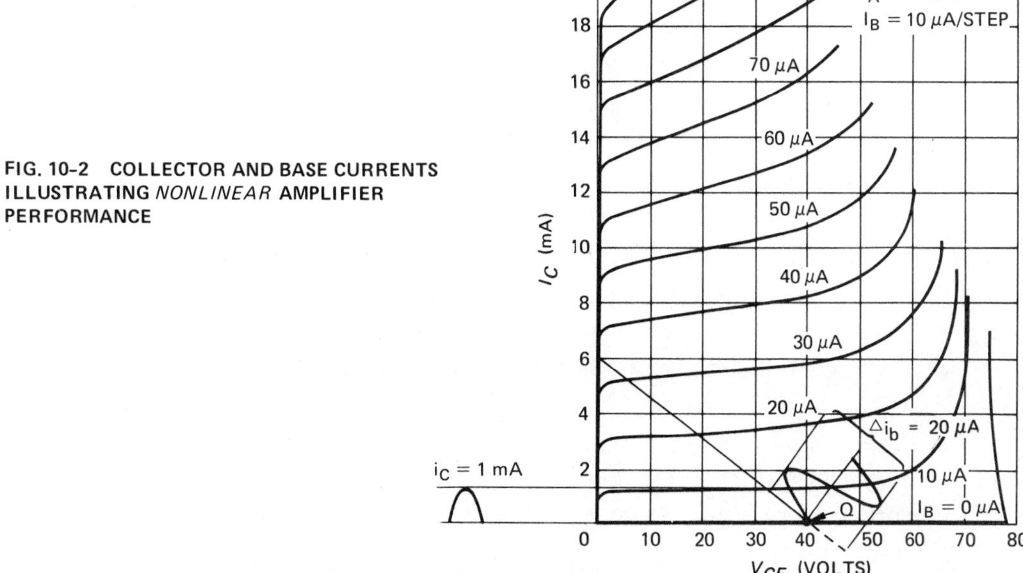

FIG. 10-2 COLLECTOR AND BASE CURRENTS ILLUSTRATING *NONLINEAR* AMPLIFIER PERFORMANCE

The base current is a sine wave that varies along the load line from the Q point at 0μA to a maximum of 10 μA and a minimun of -10 μA. The minimum value is 10 μA below the cutoff point. When the base current is 10 μA, the collector current is 1 mA. When the base current is less than 0 μA, the collector current is 0 mA for every base current value. In other words, the collector current represents only half of the sine wave. This means that the collector conducts current during the positive half cycle of the base current sine wave. The amplifier circuit, therefore, is nonlinear.

A knowledge of transistor amplifier circuit biasing is important to the electronics technician because it is very helpful in understanding circuit operation, servicing, and the effect of biasing circuits for ac operation. The ability to determine voltages and currents in transistor circuits is used when analyzing the operation of various integrated circuits. When servicing equipment, the technician will use schematic diagrams which usually show the operating conditions of the circuit. By measuring the actual voltage values, the technicians can determine the parts of a circuit that are not functioning correctly.

To bias a circuit properly, it may be necessary to sacrifice some of the ac performance quantities. For example, part of the ac power in the biasing resistors may be lost. Since part of the input and/or output current must flow in the biasing components, the ac voltage gain or current gain (or both) may be reduced. In addition, the input resistance of the circuit may be reduced if the biasing components are in parallel with the transistor.

There are three limitations to be observed when locating the operating (Q) point in the active region. These limitations are shown in figure 10-3. The current $I_{C(MAX)}$

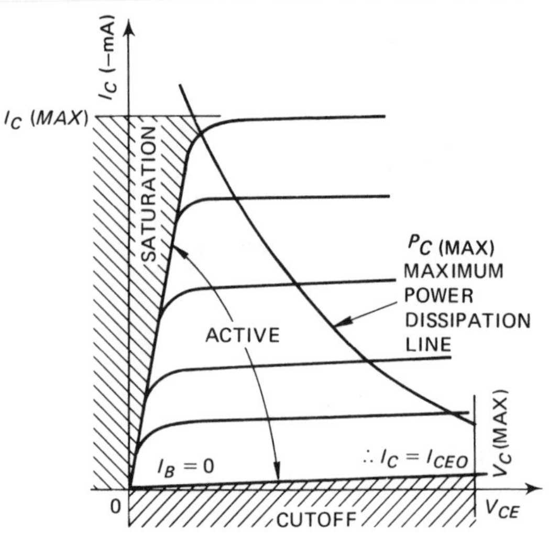

FIG. 10-3 COMMON EMITTER COLLECTOR CHARACTERISTICS

is the maximum dc saturation collector current that can flow without destroying the transistor. The voltage $V_{C(MAX)}$ is the maximum collector voltage permissible. The power $P_{C(MAX)}$ is the maximum permissible power dissipation at a particular temperature. $P_{C(MAX)}$ is a function of the collector current and voltage and thus appears as the maximum power dissipation line shown in figure 10-3.

Many types of circuits can be used to provide bias for transistor amplifiers. This unit and Unit 11 will cover those circuits used to bias transistors for CE operation. The CE bias circuits are: (a) the fixed-bias circuit, (b) the emitter-bias circuit with single base resistor, (c) the emitter-bias circuit with voltage divider, (d) the collector-base bias, (e) the collector-base bias circuit with emitter resistor, and (f) the emitter-bias circuit with two supplies.

All of the bias circuits to be analyzed use NPN transistors. If PNP transistors are substituted, all of the power supply polarities must be reversed.

To develop the equations in this unit and in Unit 11, Ohm's Law and Kirchhoff's voltage and current laws are used. The reader is encouraged *not* to memorize the equations developed. All of these equations can be obtained by examining the circuit and then applying either Ohm's Law or the Kirchhoff voltage and current laws. Once the equations are developed, they can be solved for the unknown quantities. This method of solution helps the reader become familiar with actual circuits, the use of basic dc circuit laws, and provides practice in *thinking* through a solution.

FIXED – BIAS CIRCUIT

A transistor is connected in a *fixed-bias circuit* in figure 10-4a. The name of the circuit is due to the fact that the base current I_B remains relatively constant (fixed) regardless of variations in the collector current I_C. The dc circuit is drawn as shown in figure 10-4b so that the biasing circuitry can be studied. The dc circuit retains only that portion of the original circuit in which the dc current can flow (remember that capacitors do not pass dc current).

Note that the power supply in the circuit of figure 10-4a is shown as a single battery, with the positive terminal connected to the ends of resistors R_C and R_B. The negative terminal of the supply is connected to ground. It is common practice to show the supply connected as in figure 10-4b, page 172. Here, the tops of resistors R_B and R_C are connected to $+V_{CC}$. It is understood that the other end of the supply is connected to ground.

Kirchhoff's voltage law can be applied around the supply-base-ground circuit in figure 10-4b to yield Eq. 10-1.

$$V_{CC} = I_B R_B + V_{BE} \qquad \textbf{Eq. 10.1}$$

With the exception of V_{BE}, most of the supply voltage is dropped across R_B. Since the supply does not change, I_B does not change.

Eq. 10.1 can be rewritten in terms of I_B:

$$I_B = \frac{V_{CC} - V_{BE}}{R_B} \qquad \textbf{Eq. 10.2}$$

V_{CC} is usually much greater than V_{BE}; therefore, V_{BE} can be neglected in Eq. 10.2.

Applying Kirchhoff's voltage law around the supply-collector-ground circuit yields Eq. 10.3.

$$V_{CC} = I_C R_C + V_{CE} \qquad \textbf{Eq. 10.3}$$

When a fixed-bias transistor circuit, as shown in figure 10-4, is designed for a specific operating point, Eq. 10.1 and Eq. 10.3 are used to solve for the values of resistors R_B and R_C. Problem 1 illustrates steps in designing a fixed-bias transistor circuit.

The fixed-bias transistor circuit is widely used as a *solid-state switch.* The transistor is switched from the saturation region (ON position) to the cutoff region (OFF position) by the application of a specific base current. When the transistor is biased in the saturation region, care must be taken in selecting the base biasing resistor

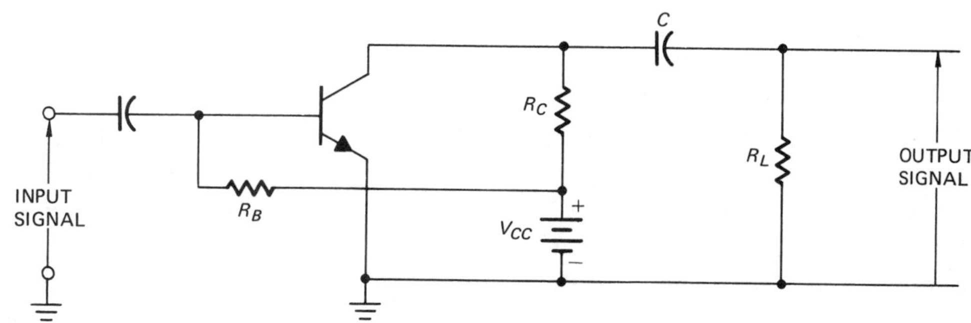

FIG. 10-4a TRANSISTOR CONNECTED IN A FIXED-BIAS CIRCUIT

R_B compared to the collector biasing resistor R_C. If a transistor is saturated, the following parameters are known.

$V_{CES} = 0$ (the subscript S means saturation)

$$I_{CS} = \frac{V_{CC}}{R_C} \qquad \textbf{Eq. 5.7}$$

$$I_{CS} = \beta I_{BS} \qquad \textbf{Eq. 7.7}$$

From Eq. 10.2, if $V_{BE} = 0$:

$$I_{BS} = \frac{V_{CC}}{R_B}$$

If this relationship for I_{BS} is substituted into Eq. 7.7 and then Eq. 5.7 is equated to Eq. 7.7, an expression for R_B is obtained.

$$\frac{V_{CC}}{R_C} = \frac{\beta V_{CC}}{R_B}$$

$$R_B = \beta R_C \qquad \textbf{Eq. 10.4}$$

Eq. 10.4 indicates that the value of the base-biasing resistor is equal to β times the collector-biasing resistor when a transistor operates in the saturation region. Eq. 10.4 also *implies* that the value of the base-biasing resistor may be slightly less than β times the collector-biasing resistor. This condition permits *slightly more base current* to flow than is necessary to achieve saturation region operation. As a result, saturation is assured. R_B must be selected carefully since an excessive amount of base current may destroy the transistor.

PROBLEM 1.

An NPN silicon transistor in the circuit in figure 10-4b has an h_{FE} (or β) equal to 100 and negligible I_{CBO}. Using a 20-V supply, design a fixed-bias common-emitter circuit for an operating point of $I_C = 1.0$ mA and $V_{CE} =$ 10 V.

For a silicon transistor, $V_{BE} = 0.6$ V. Applying Kirchhoff's voltage law around the supply-collector-emitter-ground circuit yields:

$$V_{CC} = I_C R_C + V_{CE}$$

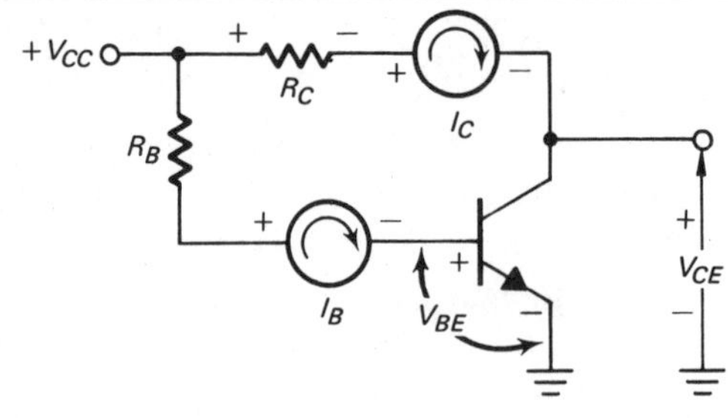

FIG. 10-4b DC CIRCUIT OF FIG. 10-4a. REMEMBER THAT THE OTHER END OF THE V_{CC} SUPPLY IS CONNECTED TO GROUND

$$R_C = \frac{V_{CC} - V_{CE}}{I_C} = \frac{20\text{ V} - 10\text{ V}}{1\text{ mA}} = \mathbf{10\ k\Omega}$$

Applying Kirchhoff's voltage law around the supply-base-resistor-base-emitter-ground the circuit yields:

$$V_{CC} = I_B R_B + V_{BE}$$

In the active region with $I_{CBO} = 0$, use Eq. 7.7 to find I_B.

$$I_B = \frac{I_C}{\beta} = \frac{1\text{ mA}}{100} = 10\ \mu\text{A}$$

Therefore,

$$R_B = \frac{V_{CC} - V_{BE}}{I_B} = \frac{20\text{ V} - 0.6\text{ V}}{10\ \mu\text{A}} = \frac{19.4\text{ V}}{10\ \mu\text{A}} = \mathbf{1.94\ M\Omega}$$

PROBLEM 2.

For the CE fixed-bias circuit shown in figure 10-5, determine the operating point.

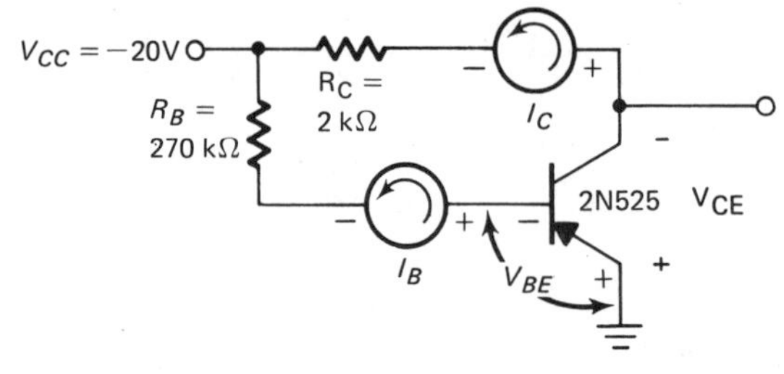

FIG. 10-5

Determine β.

The typical value for β listed on the 2N525 transistor specification sheet is 50. Note the operating point at which β was determined: I_C = -20 mA and V_{CE} = -1 V. If collector characteristics are not available, then this value for β must be used. Since the collector characteristics are available, figure 10-6, a more accurate value of β can be determined.

First, the load line is drawn for the given circuit.

Point 1: Let $I_C = 0$, then $V_{CE} = V_{CC}$ = -20 V.

Point 2: Let $V_{CC} = 0$, then $I_C = V_{CC}/R_C$ = -20 V/2 k Ω = -10 mA

The load line is drawn by connecting the two points just determined, figure 10-6. To insure the *linear* operation of the amplifier, assume a Q point in the middle of the load line. This means that I_{CQ} = -10 mA/2 = -5 mA, and V_{CEQ} = -10 V.

Calculate β from the operating point on the load line.

$$\beta = \frac{I_C}{I_B}$$

I_C = 5 mA from the load line in figure 10-6

I_B = 0.07 mA from the load line in figure 10-6

$$\beta \doteq \frac{5.0 \text{ mA}}{0.07 \text{ mA}} = \mathbf{71.4}$$

After the operating point values are calculated, it may be necessary to change the assumptions concerning the operating point, so that a value of β closer to the actual operating point can be determined. To find the actual value of I_B, Kirchhoff's voltage law is applied around the supply-base-emitter circuit. Since a germanium 2N525 transistor is used, V_{BE} = 0.2 V.

$$V_{CC} = I_B R_B + V_{BE}$$

$$20 \text{ V} = I_B (270 \text{ k}\Omega) + 0.2 \text{ V}$$

$$I_B = \mathbf{0.0734 \text{ mA}}$$

To find I_C, use Eq. 7.7:

$$I_C = \beta I_B = 71.4 \times 0.0734 \text{ mA}$$

$$= \mathbf{5.25 \text{ mA}}$$

To find V_{CE}, apply Kirchhoff's voltage law around the supply-collector-emitter circuit.

$$V_{CC} = I_C R_C + V_{CE}$$

$$V_{CE} = 20 \text{ V} - 5.25 \text{ mA} \times 2.0 \text{ k}\Omega$$

$$= 20 \text{ V} - 10.5 \text{ V}$$

$$= \mathbf{9.5 \text{ V}}$$

The polarities of the voltages and directions of the currents are shown in figure 10-5.

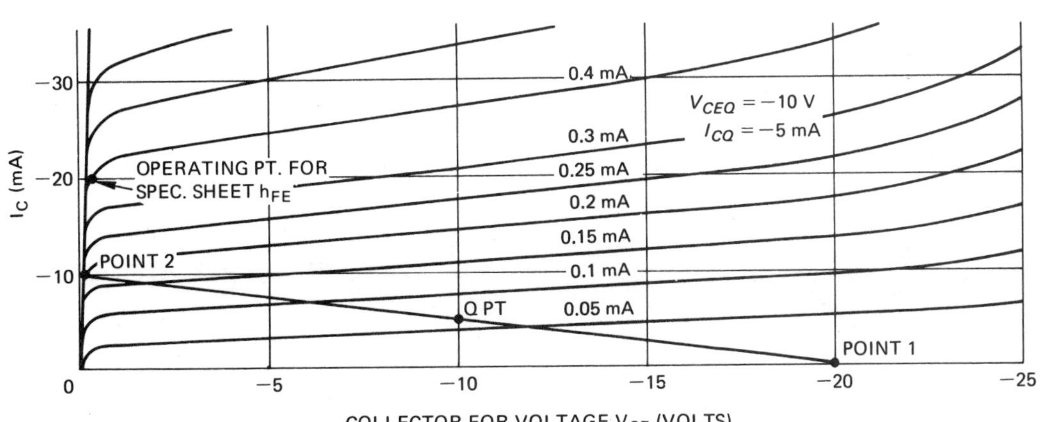

FIG. 10-6 COMMON-EMITTER COLLECTOR CHARACTERISTICS,TYPE 2N525

EMITTER-BIAS CIRCUIT WITH SINGLE BASE RESISTOR

In the fixed-bias circuit, the base current remains constant regardless of any changes in the collector current. However, it is also important that when a transistor is biased, the biasing should *stabilize the operating point.* Stability is the ability of a circuit to keep the operating point at the desired values. The stability of a fixed-bias circuit can be improved by adding a bias resistor to the emitter circuit as shown in figure 10-7.

The emitter-bias circuit is also known as a dc current feedback circuit. This is due to the fact that the current I_E flowing through the emitter resistor R_E is equal to $I_B + I_C$. The current I_C is brought from the output circuit back to the input circuit. This current helps to stabilize the circuit operation.

When Kirchhoff's voltage law is applied around the supply-collector-emitter-ground circuit in figure 10-7, the result is:

$$V_{CC} = I_C R_C + V_{CE} + I_E R_E$$

Substituting the transistor current equation into the above equation yields:

$$V_{CC} = I_C R_C + V_{CE} + (I_C + I_B) R_E$$

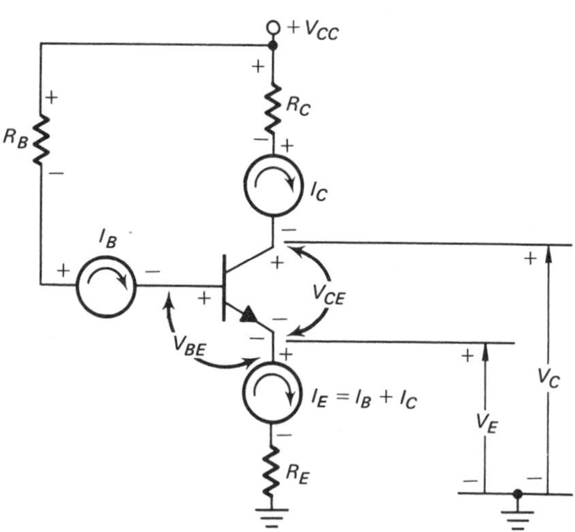

FIG. 10-7 DC CIRCUIT OF EMITTER-BIAS WITH SINGLE BASE RESISTOR

Assume that $I_{CBO} = 0$ in the active region. Substitute Eq. 2.16 into the above equation to obtain:

$$V_{CC} = I_C R_C + V_{CE} + \left(I_C + \frac{I_C}{\beta}\right) R_E$$

Solving for the collector current I_C:

$$I_C = \frac{V_{CC} - V_{CE}}{R_C + \left(\frac{\beta + 1}{\beta}\right) R_E} \qquad \text{Eq. 10.5}$$

Another method of finding the collector current is to apply Kirchhoff's voltage law around the supply-base-emitter-ground circuit.

$$V_{CC} = I_B R_B + V_{BE} + I_E R_E$$

Substituting the transistor current equation and Eq. 7.7 into the above equation yields:

$$V_{CC} = \left(\frac{I_C}{\beta}\right) R_B + V_{BE} + \left(I_C + \frac{I_C}{\beta}\right) R_E$$

Solving for the collector current I_C:

$$I_C = \frac{V_{CC} - V_{BE}}{\frac{R_B}{\beta} + \frac{(\beta + 1)}{\beta} R_E} \; R_E \qquad \text{Eq. 10.6}$$

Equation 10.6 can be written in terms of the base current by substituting Eq. 7.7 (when this equation is written in terms of the collector current).

$$I_B = \frac{V_{CC} - V_{BE}}{R_B + (\beta + 1)\ R_E} \qquad \text{Eq. 10.7}$$

To solve Eq. 10.5, the voltage V_{CE} must be known. This voltage is simply the collector voltage minus the emitter voltage.

$$V_{CE} = V_C - V_E$$

$$V_C = V_{CC} - I_C R_C \qquad \text{Eq. 10.8}$$

$$V_E = I_E R_E = (I_B + I_C) R_E = \left(I_C + \frac{I_C}{\beta}\right) R_E$$

$$V_E = \frac{(\beta + 1)}{\beta} I_C R_E \qquad \text{Eq. 10.9}$$

An expression for V_{CE} can be written by subtracting Eq. 10.9 from Eq. 10.8:

$$V_{CE} = V_{CC} - I_C R_C - \frac{(\beta + 1)}{\beta} I_C R_E \qquad \text{Eq. 10.10}$$

The equations just developed can now be used to determine the values of the biasing resistors for a given operating point for the circuit in figure 10-7.

PROBLEM 3.

The circuit in figure 10-5 contains an NPN silicon transistor with β equal to 100 and negligible I_{CBO}. Using a 20-V power supply, design an emitter bias circuit with single base resistor for an operating point of I_C = 1.0 mA and V_{CE} = 10 V.

For a silicon transistor, $V_{BE} = 0.6$ V

In the active region when $I_{CBO} = 0$,

$$I_B = \frac{I_C}{\beta} = \frac{1 \text{ mA}}{100} = 10\ \mu\text{A}$$

Applying Kirchhoff's voltage law around the collector-emitter circuit in figure 10-7 yields:

$$V_{CC} = I_C R_C + V_{CE} + I_E R_E$$

$$20 = 10^{-3} \times R_C + 10 + 1.01 \times 10^{-3} \times R_E$$

From the Kirchhoff voltage law equation around the base-emitter circuit:

$$V_{CC} = I_B R_B + V_{BE} + I_E R_E$$

$$20 = 10 \times 10^{-6} \times R_B + 0.6 + 1.01 \times 10^{-3} \times R_E$$

At this point, it is necessary to solve two equations with three unknowns, R_B, R_C, and R_E. Therefore, a value is selected for one of the unknown resistors so that the two other resistor values can be determined. A value for R_E is usually selected. R_E is a *swamping* resistor; that is, it is placed in the circuit to help make I_C independent of β. As a result, R_E should be as large as possible. The following approximations are made so that an approximate value of R_E can be determined.

$$I_E = I_C = 1 \text{ mA, when } V_{CE} = 10 \text{ V}$$

For the Kirchhoff voltage law equation for the circuit in figure 10-7,

$$V_{CC} = V_{CE} + I_C\ (R_C + R_E)$$

$$R_C + R_E = \frac{V_{CC} - V_{CE}}{I_C} = \frac{(20 - 10)\text{ V}}{1 \text{ mA}}$$

$$= 10 \text{ k}\Omega$$

The equation $R_C + R_E = 10 \text{ k}\Omega$ puts critical restrictions on R_E. Assuming a value of 2 kΩ for R_E, the previous equations can be solved for R_C and R_B.

$$R_C = \frac{(20 - 10 - 1.01 \times 10^{-3} \times 2 \times 10^{+3})\text{ V}}{10^{-3} \text{ A}}$$

$$= \frac{7.98 \text{ V}}{10^{-3} \text{ A}} = 7.98 \text{ k}\Omega$$

$$R_B = \frac{(20 - 0.6 - 1.01 \times 10^{-3} \times 2 \times 10^{+3}}{10 \times 10^{-6} \text{ A}} \text{ V}$$

$$= \frac{17.38 \text{ V}}{10 \times 10^{-6} \text{ A}} = \mathbf{1.738\ M\Omega}$$

Problem 1 and Problem 3 both have large values for R_B. Not only are resistors of this size difficult to obtain, but they tend to make the circuit unstable. If the current through R_B changes slightly, then a significant voltage change will take place because of the large value of the resistor. One method of overcoming this problem is to replace R_B with a voltage divider. An emitter-bias circuit with a voltage divider is covered in Unit 11.

LABORATORY EXERCISE 10-1: FIXED-BIAS CIRCUIT

PURPOSE

- To design a fixed-bias CE circuit for a specified operating point.
- To construct a fixed-bias circuit in the laboratory and measure the β of the transistor and the circuit operating point.

- To observe the variation of the operating point with changes in the value of the base resistor.
- To compare actual laboratory data with the theoretical calculations.

MATERIALS

1 NPN transistor, 2N4425, 2N3405 or equivalent
1 Dc power supply, 10 V
1 VTVM, or solid-state voltmeter
2 VOMs, 20,000 ohms per volt
Miscellaneous ½-W resistors to be determined in step A of the Procedure

PROCEDURE

A. 1. Draw a fixed-bias CE circuit using an NPN transistor and label all components. Insert ammeters to measure I_B and I_C and show the voltages V_{BE} and V_{CE}. Using a 10-V supply, take readings and perform the necessary calculations to establish an operating point of $I_C = 10$ mA and $V_{CE} = 5$ V. Assume $V_{BE} = 0.6$ V and I_{CBO} is negligible.

2. Determine R_C. Record this value in Table 10-1 for step A.2.

TABLE 10-1

Procedure	Comments	β	R_C in Ω	R_B in Ω	I_C in mA	V_{CE} in V	I_B in μA	V_{BE} in V
A.2	Theoretical Calculation	330			10	5		0.6
A.3	Procedure	330						
A.3	Procedure	250						
A.3	Procedure	150						
A.4	Saturation	—						
C	Procedure							
D	Vary R_B							
E	Saturation	—						

3. Determine and record the values of I_B and R_B for each value of β in Table 10-1 for step A.3.

4. Determine the values of I_{BS} and R_B required for saturation. To determine the value of R_B, use the minimum $\beta = 150$. Record the calculated values in step A.4. of Table 10-1.

B. Connect the circuit drawn in step A.1. Use the values of V_{CC} and R_C as determined or given in step A. Assume that $\beta = 330$ and use the value determined for R_B in step A.3. Use resistors with values as close to the calculated values as possible. Two resistors connected in series or in parallel may be used if necessary. *Do not use potentiometers.* List the values of the resistors used in step C. of Table 10-1.

C. Complete all of the data required for step C. of Table 10-1. Determine the actual value of β for the transistor using the measured values for I_C and I_B.

D. For the operating point obtained in step C, if V_{CE} is not greater than 4 V and less than 6 V, change the base resistor so that V_{CE} is approximately 5 V. Complete step D. in Table 10-1.

E. If step D was performed, disregard step E. and go to step F. If V_{CE} for the operating point is between 4 V and 6 V, change R_B to the value obtained in step A.3. for $\beta = 150$ to observe the effect of changing resistance on the operating point. Complete step D in Table 10-1.

F. Change the value of R_B to that determined for saturation and complete step E of Table 10-1.

DISCUSSION QUESTIONS

1. Compare the theoretical and actual test results of steps A.2 and C in Table 10-1. Discuss the results and give reasons for any discrepancies.

2. Does the transistor saturate in step E? Discuss the test results. Why is the minimum value of β used in the calculation for R_B?

3. Using the measured values of step C, verify Kirchhoff's voltage law around the base-emitter circuit loop and the collector-emitter circuit loop.

4. In the fixed-bias CE amplifier circuit, does the operating point depend upon the β of the transistor? Discuss and use laboratory data to substantiate this answer.

LABORATORY EXERCISE 10-2: EMITTER-BIAS CIRCUIT

PURPOSE

- To design an emitter-bias CE circuit with a single base resistor for a specific operating point.
- To construct an emitter-bias circuit in the laboratory and measure the β of the transistor and the circuit operating point.
- To compare actual laboratory data with the theoretical calculations.

MATERIALS

1 NPN transistor, 2N4425, 2N3405 or equivalent
1 Dc power supply, 10 V
1 VTVM, or solid-state voltmeter
2 VOMs, 20,000 ohms per volt
Miscellaneous ½-W resistors to be determined in step A of the Procedure

PROCEDURE

A. 1. Draw an emitter-bias CE circuit with a single base resistor using an NPN transistor. Label all components. Insert ammeters to measure I_B and I_C. Show the voltages V_C, V_E, V_{CE} and V_{BE}. Using a 10-V supply with $R_E = 200\ \Omega$, perform the necessary calculations to establish an operating point of $I_C = 10$ mA and $V_{CE} = 5$ V. Assume that $V_{BE} = 0.6$ V and I_{CBO} is negligible.

2. Determine R_C. Record this value for step A.2. in Table 10-2.

3. Determine the values of I_B and R_B for each value of β given in Table 10-2 for steps A.2. and A.3. Record the values of I_B and R_B at $\beta = 330$ for step A.2. of Table 10-2. Record the remaining calculations in Table 10-2, step A.3.

B. 1. Connect the circuit drawn in step A.1. Use the values of V_{CC}, R_E, and R_C as given or determined in step A. Assume that $\beta = 330$ and use the value determined for R_B in step A.2. Use resistor values as close to the calculated values as possible. Two resistors connected in series or in parallel may be used if necessary. *Do not use potentiometers.*

2. Record the values of the resistors used in step C of Table 10-2.

C. Determine and record all of the data required for step C of Table 10-2. Determine the actual value of β for the transistor using the measured values for I_C and I_B.

D. For the operating point obtained in step C, if V_{CE} is not greater than 4 V and less than 6 V, change the base resistor so that V_{CE} is approximately 5 V. Complete step D in Table 10-2.

E. If step D was performed, disregard step E. If V_{CE} at the operating point is between 4 V and 6 V, change R_B to the value found in step A.3. for $\beta = 150$ to observe the effect of changing resistance on the operating point. Complete step D in Table 10-2.

TABLE 10-2

Procedure Step	Comments	β	R_C in Ω	R_E in Ω	R_B in Ω	I_C in mA	V_C in V	V_E in V	V_{CE} in V	I_B in μA	V_{BE} in V
A.2	Theoretical Calculation	330		200		10	7	2	5		0.6
A.3	Procedure	250									
A.3	Procedure	150									
C	Actual Lab Data										
D	Vary R_B										

DISCUSSION QUESTIONS

1. Compare the theoretical and actual test results of steps A.2. and C. in Table 10-2. Discuss the results and give reasons for any discrepancies.

2. Using the measured values of step C., verify Kirchhoff's voltage law around the base-emitter circuit loop and the collector-emitter circuit loop.

3. In the emitter-bias CE amplifier circuit, does the operating point depend upon the β of the transistor? Discuss and use laboratory data to substantiate this answer.

4. Compare the results of the fixed-bias and emitter-bias CE amplifier circuits. For which circuit is the actual operating point closer to the theoretical operating point? Why?

5. Which circuit is *less* β dependent? Why?

EXTENDED STUDY TOPICS

1. The circuit in figure 10-8 is a fixed-bias PNP or NPN transistor. Refer to figure 10-8 and find the missing quantities in all parts of Table 10-3.

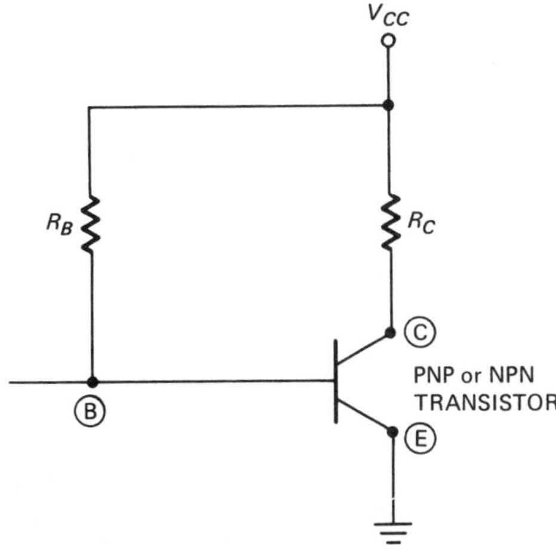

FIG. 10-8

TABLE 10-3

	Type	β	V_{BE} in V	V_{CC} in V	R_C in k Ω	R_B in MΩ	I_{CBO} in μA	I_C in mA	I_B in μA	V_{CE} in V
a	PNP	49	0.2	-20			3	10		-10
b	NPN	200	0.6	+22	5		0			+12
c	NPN	250	0.6	+15			0	1		+5
d	NPN	100	0.6	+15	10		0		10	
e	PNP	100	0.2	-15	5	1.48	0			
f	NPN	40	0.6	+15			0	1		+5
g	NPN	200	0.6	+15			1	1		+5
h	NPN	40	0.6	+20	10		5	1		
i	PNP	39	0.2	-20			5	3		-6

2. The circuit in figure 10-9 is an emitter-bias circuit with a single base resistor and a PNP or NPN transistor. Find the missing quantities in Table 10-3 for the following:

   a. Table 10-3, part a, with $R_E = 1\ k\,\Omega$

   b. Table 10-3, part c, with $R_E = 2\ k\,\Omega$

   c. Table 10-3, part f, with $R_E = 1\ k\,\Omega$

   d. Table 10-3, part i, with $R_E = 500\ \Omega$

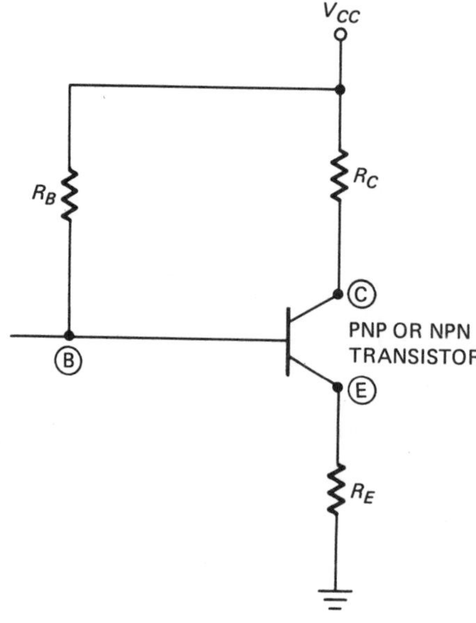

FIG. 10-9

More common-emitter biasing

OBJECTIVES

After studying this unit, the student will be able to discuss and demonstrate an understanding of the basic principles of:

- Analyzing the following common-emitter bias circuits:

  Emitter-bias circuit with voltage divider
  Collector-base bias circuit
  Collector-base bias circuit with emitter resistor
  Emitter-bias circuit with two power supplies

EMITTER-BIAS CIRCUIT WITH VOLTAGE DIVIDER

The emitter-bias circuit with a voltage divider is shown in figure 11-1. The circuit appears to be difficult to analyze because the current in the base circuit consists of current through resistors R_1 and R_2 plus the base current. However, the circuit can be simplified using Thevenin's Equivalent Circuit. The part of the circuit that is treated as the load in Thevenin's Equivalent Circuit is shown enclosed by a dotted line in figure 11-2a, page 182.

To obtain the open circuit voltage source, V_{OC}, in Thevenin's circuit in figure 11-2, the load is removed and the open circuit voltage is determined across points A and B in figure 11-2a. Since the resistors R_1 and

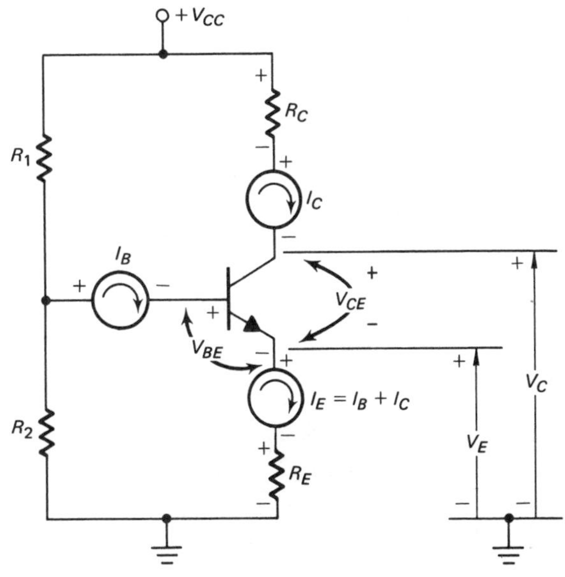

FIG. 11-1 EMITTER-BIAS CIRCUIT WITH VOLTAGE DIVIDER

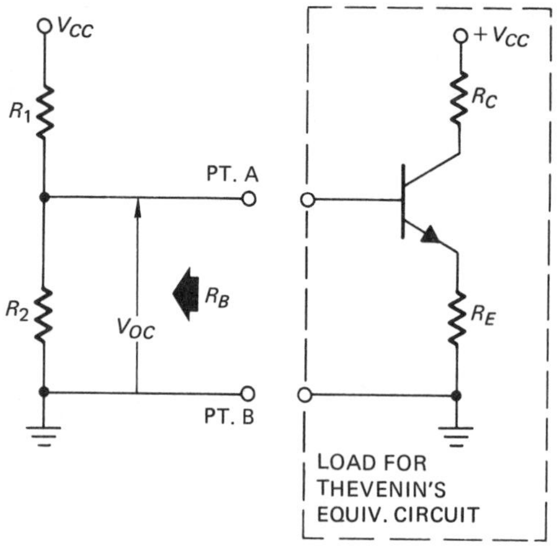

(a) CIRCUIT SHOWING LOAD FOR THEVENIN'S EQUIVALENT CIRCUIT

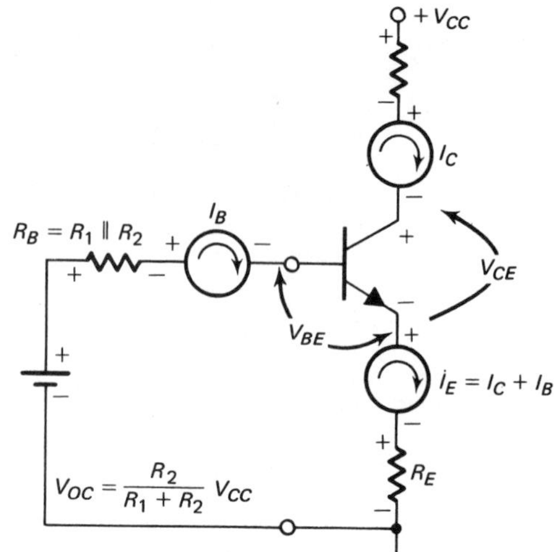

(b) THEVENIN'S EQUIVALENT CIRCUIT FOR (a) WITH LOAD ATTACHED

FIG. 11-2 EMITTER-BIAS CIRCUIT WITH VOLTAGE DIVIDER EQUIVALENT CIRCUIT

R_2 form a voltage divider, the open circuit voltage is found using the equation:

$$V_{OC} = \frac{R_2}{R_1 + R_2} V_{CC}$$

Thevenin's resistance R_B in figure 11-2b is determined by finding the resistance to the left of points A and B in figure 11-2a after all of the sources are replaced by their internal resistances.

$$R_B = R_1 \| R_2 = \frac{R_1 R_2}{R_1 + R_2}$$

Thevenin's circuit in figure 11-2b is used to develop the expressions that determine the bias conditions. Applying Kirchhoff's voltage law around the base circuit yields:

$$V_{OC} = I_B R_B + V_{BE} + (I_C + I_B) R_E$$

In the active region, assume that $I_{CBO} = 0$. Substitute Eq. 7.7 into the equation for V_{OC} and solve for the base current, I_B.

$$V_{OC} = I_B R_B + V_{BE} + (\beta I_B + I_B) R_E$$

$$\text{Thus, } I_B = \frac{V_{OC} - V_{BE}}{R_B + (\beta + 1) R_E} \qquad \text{Eq. 11.1}$$

To determine the collector current, an expression is obtained for Kirchhoff's voltage law around the collector circuit. The transistor current equation for I_E is then substituted into the Kirchhoff expression.

$$V_{CC} = I_C R_C + V_{CE} + (I_C + I_B) R_E$$

Substituting Eq. 7.7 (expressed in terms of I_B) into the above equation, solve for the collector current.

$$V_{CC} = I_C R_C + V_{CE} + (I_C + \frac{I_C}{\beta}) R_E$$

$$I_C = \frac{V_{CC} - V_{CE}}{R_C + \frac{(\beta + 1)}{\beta} R_E} \qquad \text{Eq. 11.2}$$

The following equations express the transistor voltages.

$$V_C = V_{CC} - I_C R_C \qquad \text{Eq. 11.3}$$

$$V_{CE} = V_C - V_E \qquad \text{Eq. 11.4}$$

$$V_E = I_E R_E = (I_C + I_B) R_E = (I_C + \frac{I_C}{\beta}) R_E$$

$$V_E = \frac{(\beta + 1)}{\beta} I_C R_E \qquad \text{Eq. 11.5}$$

If the voltage divider circuit has *no* emitter resistor, then these voltage equations apply if R_E is set equal to zero. It will be shown in a more advanced course that R_B should be made ten (10) times larger than R_E in figure 11-2 to achieve good stability.

Once the value for R_B is known, R_1 and R_2 can be determined as shown in the following procedure.

$$V_{OC} = \left[\frac{R_2}{R_1 + R_2}\right] V_{CC}$$

$$R_B = \frac{R_1 \; R_2}{R_1 + R_2}$$

Solving for R_B/R_1:

$$\frac{R_B}{R_1} = \frac{R_2}{R_1 + R_2}$$

Substituting the expression for R_B/R_1 into the expression for V_{OC} gives the following equation.

$$V_{OC} = \frac{R_B}{R_1} V_{CC}$$

Solving for R_1:

$$R_1 = R_B \frac{V_{CC}}{V_{OC}} \qquad \text{Eq. 11.6}$$

By repeating this procedure for R_B/R_2 and solving for R_2, an expression for R_2 is obtained.

$$R_2 = R_B \left[\frac{V_{CC}}{V_{CC} - V_{OC}}\right] \qquad \text{Eq. 11.7}$$

PROBLEM 1.

The circuit in figure 11-1 contains an NPN silicon transistor with β equal to 100 and negligible I_{CBO}. Using a 20-V supply, design an emitter-bias circuit with a voltage divider for an operating point of I_C = 1.0 mA and V_{CE} = 10 V. Assume that R_E = 2 k Ω and R_B = 10 R_E (to provide stability).

For a silicon transistor, V_{BE} = 0.6 V.

As in Problem 3, Unit 10, it is necessary to solve two equations with three unknowns: R_B, R_E, and R_C. Therefore, a value must be assigned to one of the resistors. A later course in electronic amplifiers will illustrate methods of assigning a value for R_B or R_E depending upon the stability of the circuit. For this problem, R_E is assigned a value of 2 k Ω.

In the active region, when I_{CBO} = 0,

$$I_B = \frac{I_C}{\beta} = \frac{1 \text{ mA}}{100} = 10 \mu\text{A}$$

Applying Kirchhoff's voltage law around the collector circuit yields:

$$V_{CC} = I_C R_C + V_{CE} + (I_C + I_B) R_E$$

$$R_C = \frac{V_{CC} - V_{CE} - (I_C + I_B) R_E}{I_C}$$

$$= \frac{20 - 10 - 1.01 \times 10^{-3} \times 2 \times 10^{+3}}{1 \times 10^{-3}}$$

$$= 7.98 \text{ k}\Omega$$

Applying Kirchhoff's voltage law around the base circuit yields:

$$V_{OC} = I_B R_B + V_{BE} + (I_C + I_B) R_E$$

Again, there are too many unknowns. Therefore, the stability approximation R_B = 10 R_E is used so that R_B equals 20 k Ω.

$$V_{OC} = 1 \times 10^{-5} \times 2 \times 10^{+4} + 0.6 + 1.01 \times 10^{-3} \times 2 \times 10^{+3} = 2.82 \text{ V}$$

Solving for R_1 and R_2:

$$R_1 = R_B \frac{V_{CC}}{V_{OC}} = \frac{2 \times 10^{+4} \times 20}{2.82}$$

$$= 142 \text{ k}\Omega$$

$$R_2 = \left[\frac{R_B \; V_{CC}}{V_{CC} - V_{OC}}\right] = 20 \times 10^{+4}\left[\frac{20}{20 - 2.82}\right]$$

$$= 23.3 \text{ k}\Omega$$

PROBLEM 2.

Determine the operating point of the circuit shown in figure 11-3.

Draw Thevenin's equivalent circuit as shown in figure 11-4. Solve for the open circuit voltage source V_{OC} and the Thevenin resistance R_B.

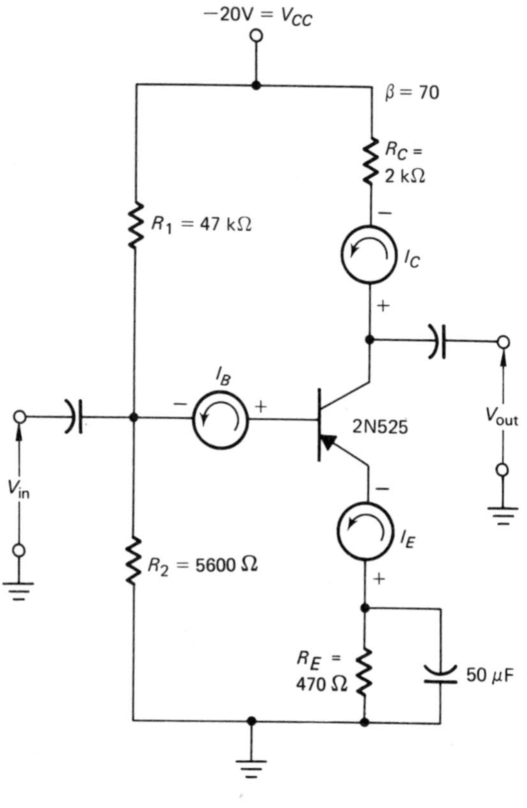

FIG. 11-3

$$V_{OC} = \left[\frac{R_2}{R_1 + R_2}\right] V_{CC} = \left[\frac{5.6\text{ k}\Omega}{5.6\text{k}\Omega + 47\text{k}\Omega}\right] 20\text{ V}$$

$$= \mathbf{2.13\ V}$$

$$R_B = R_1 \parallel R_2 = \frac{5.6\text{ k}\Omega \times 47\text{ k}\Omega}{5.6\text{ k}\Omega + 47\text{ k}\Omega}$$

$$= 5\text{ k}\Omega = \mathbf{5000\ \Omega}$$

To solve for base current, use Eq. 11.1 and V_{BE} = 0.2 V (for the 2N525 germanium transistor).

$$I_B = \frac{V_{OC} - V_{BE}}{R_B + (\beta + 1)\, R_E}$$

$$= \frac{2.13\text{ V} - 0.2\text{ V}}{5000\ \Omega + 71 \times 470\ \Omega}$$

$$= \mathbf{0.0502\ mA}$$

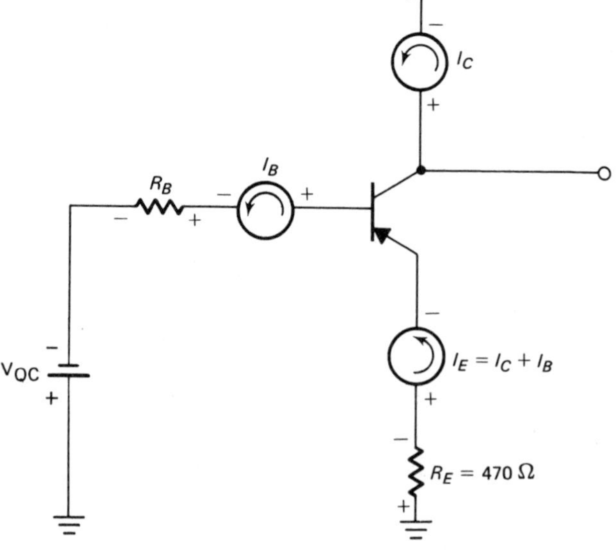

FIG. 11-4 DC THEVENIN'S EQUIVALENT CIRCUIT FOR FIGURE 11-3

To solve for the collector current, assume that $I_{CBO} = 0$

$$I_C = \beta I_B = 70 \times 0.0502 \text{ mA}$$

$$= \mathbf{3.52\ mA}$$

The emitter current is found by the use of the transistor current equation.

$$I_E = I_C + I_B = 3.52 \text{ mA} + 0.0502 \text{ mA}$$
$$= 3.5702 \text{ mA}$$

$$V_{CC} = I_C R_C + V_{CE} + I_E R_E$$

$$V_{CE} = V_{CC} - I_C R_C - I_E R_E$$

$$= 20 \text{ V} - 2 \times 10^3 \times 3.52 \times 10^{-3} - 470 \times 3.57 \times 10^{-3}$$

$$= \mathbf{11.28\ V}$$

Thus, the Q point is defined as:

I_{CQ} - 3.52 mA, I_B = 0.0502 mA, and V_{CEQ} = 11.28 V

Since a PNP transistor is used in the circuit, the voltages determined in this problem are actually *negative* with respect to ground. The NPN transistor voltages are *positive* with respect to ground.

For the emitter bias circuit with voltage divider shown in figure 11-5, find the missing quantities in Table 11-1. Neglect I_{CBO}. (R11-1)

COLLECTOR-BASE BIAS

The circuit in figure 11-6 improves the stability of the fixed-bias circuit. Note that R_B is not connected to the supply voltage V_{CC} as in previous circuits, but is connected to the collector of the transistor. This circuit is also known as a dc voltage feedback circuit because the output voltage is fed back to the input circuit. The collector-base resistor R_B provides negative feedback from the output circuit to the input circuit to help stabilize the circuit operation.

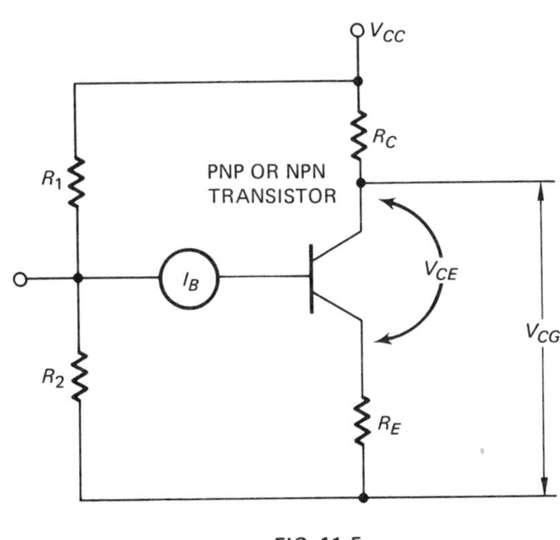

FIG. 11-5

TABLE 11-1

Type	β	V_{BE} in V	V_{CC} in V	R_1 in kΩ	R_2 in kΩ	R_C in kΩ	R_E in kΩ	I_C in mA	I_E in mA	I_B in μA	V_{CE} in V	V_{CG} in V
NPN	50	0.6	+10	55	12.0	4.0	1.0			20		

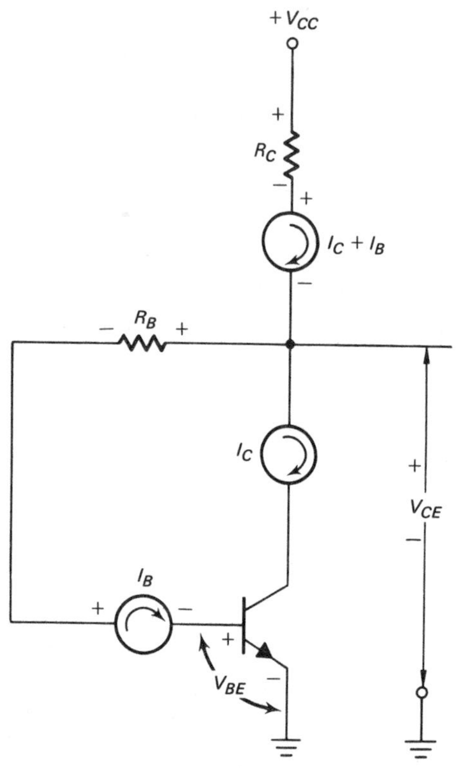

FIG. 11-6 COLLECTOR-BASE BIAS CIRCUIT

Writing Kirchhoff's voltage law around the supply-collector-resistor-base-ground circuit in figure 11-6 yields:

$$V_{CC} = (I_C + I_B)\, R_C + I_B R_B + V_{BE} \qquad \text{Eq. 11.8}$$

In the active region, assume that I_{CBO} = 0. Substitute Eq. 2.16 into Eq. 11.8 to obtain:

$$V_{CC} = (\beta\, I_B + I_B)\, R_C + I_B R_B + V_{BE} \qquad \text{Eq. 11.9}$$

Solving for base current I_B:

$$I_B = \frac{V_{CC} - V_{BE}}{(\beta + 1)\, R_C + R_B} \qquad \text{Eq. 11.10}$$

Another expression may be obtained for I_B from Ohm's Law.

$$I_B = \frac{V_{CE} - V_{BE}}{R_B} \qquad \text{Eq. 11.11}$$

To solve Eq. 11.8 for I_C, substitute Eq. 7.7 for the base current in terms of the collector current.

$$V_{CC} = (I_C + \frac{I_C}{\beta})\; R_C + \frac{I_C}{\beta}\, R_B + V_{BE}$$

$$I_C = \frac{V_{CC} - V_{BE}}{\frac{(\beta + 1)}{\beta}\, R_C + \frac{R_B}{\beta}} \qquad \text{Eq. 11.12}$$

The collector-emitter voltage may be found from the Kirchhoff's voltage equation around the supply-collector-emitter-ground circuit in figure 11-6.

$$V_{CE} = V_{CC} - (I_C + I_B)\, R_C \qquad \text{Eq. 11.13}$$

PROBLEM 3.

The circuit in figure 11-6 contains an NPN silicon transistor with β = 100 and negligible I_{CBO}. Using a 20-V supply, design a collector-base bias common emitter circuit for an operating point of I_C = 1.0 mA and V_{CE} = 10 V.

For a silicon transistor, V_{BE} = 0.6 V. In the active region, when $I_{CBO} = 0$,

$$I_B = \frac{I_C}{\beta} = \frac{1\text{ mA}}{100} = 10\,\mu\text{A}$$

The base and collector resistor values are found by using Ohm's Law.

$$R_B = \frac{V_{CE} - V_{BE}}{I_B} = \frac{10\text{ V} - 0.6\text{ V}}{10\,\mu\text{A}}$$

$$= 940\text{ k}\Omega$$

$$R_C = \frac{V_{CC} - V_{CE}}{I_C + I_B} = \frac{20\text{ V} - 10\text{ V}}{1\text{ mA} + 0.01\text{ mA}}$$

$$= \frac{10\text{ V}}{1.01\text{ mA}}$$

$$= 9.9\text{ k}\Omega$$

Design a collector-base bias common emitter transistor using a 2N4425 silicon transistor. The operating point is I_C = 5 mA and V_{CE} = 4 V, V_{CC} = 10 V, β = 330, and I_{CBO} = 0. (R11-2)

COLLECTOR-BASE BIAS CIRCUIT WITH EMITTER RESISTOR

The operation or stability of the collector-base bias circuit can be improved by the addition of an emitter resistor. The collector-base bias circuit with an emitter resistor is shown in figure 11-7. This circuit provides dc *voltage feedback* due to resistor R_B and dc *current feedback* due to resistor R_E. The circuit in figure 11-7 can be analyzed in a manner similar to that used for the collector-base bias circuit. Applying Kirchhoff's voltage law around the supply-collector resistor-base resistor-ground circuit yields:

$$V_{CC} = (I_C + I_B) R_C + I_B R_B + V_{BE} + (I_C + I_B) R_E$$

Eq. 11.14

In the active region, assume that $I_{CBO} = 0$. Substitute Eq. 7.7 into Eq. 11.14 and solve for the base current I_B:

$$I_B = \frac{V_{CC} - V_{BE}}{(\beta + 1) R_C + R_B + (\beta + 1) R_E} \quad \textbf{Eq. 11.15}$$

Eq. 11.15 can be solved for the collector current by substituting Eq. 7.7 for the base current expressed in terms of the collector current.

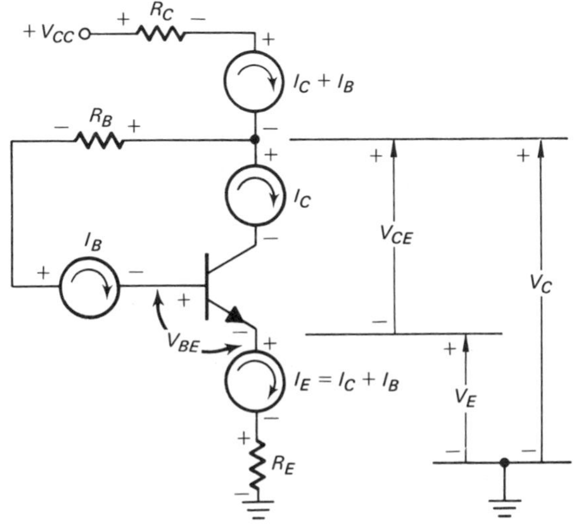

FIG. 11-7 COLLECTOR-BASE BIAS CIRCUIT WITH EMITTER RESISTOR

$$I_C = \frac{V_{CC} - V_{BE}}{\frac{(\beta + 1)}{\beta} R_C + \frac{R_B}{\beta} + \frac{(\beta + 1)}{\beta} R_E}$$

Eq. 11.16

To determine the transistor voltages, apply Kirchhoff's voltage law around the supply-collector resistor-emitter resistor-ground circuit:

$$V_{CE} = V_{CC} - (I_C + I_B) R_C - (I_C + I_B) R_E$$

Eq. 11.17

$$V_C = V_{CC} - (I_C + I_B) R_C$$

Eq. 11.18

$$V_E = (I_C + I_B) R_E$$

Eq. 11.19

PROBLEM 4.

The circuit in figure 11-7 contains an NPN silicon transistor with $\beta = 100$ and negligible I_{CBO}. Using a 20-V supply, design a collector-base bias circuit with an emitter resistor for an operating point of $I_C = 1.0$ mA and $V_{CE} = 10$ V.

As in Problem 3, unit 10 and Problem 1, unit 11, it is necessary to solve two equations with three unknowns. As in the previous problems, assume that $R_E = 2\ k\,\Omega$.

In the active region, when $I_{CBO} = 0$,

$$I_B = \frac{I_C}{\beta} = \frac{1\text{ mA}}{100} = 10\,\mu\text{A}$$

Solve for the value of the collector resistor by writing the Kirchhoff voltage expression around the collector-emitter circuit.

$$V_{CC} = (I_C + I_B) R_C + V_{CE} + (I_C + I_B) R_E$$

$$R_C = \frac{V_{CC} - V_{CE} - (I_C + I_B) R_E}{I_C + I_B}$$

$$= \frac{20 - 10 - 1.01 \times 10^{-3} \times 2 \times 10^3}{1.01 \times 10^{-3}}$$

$$= \mathbf{7.9\ k\Omega}$$

Using Ohm's Law, solve for the base resistor R_B (V_{BE} = 0.6 V for a silicon transistor).

$$R_B = \frac{V_{CE} - V_{BE}}{I_B} = \frac{10\ V - 0.6\ V}{10\mu\ A}$$

$$= 940\ k\Omega$$

Design a collector-base bias, common-emitter transistor circuit with an emitter resistor using a 2N4425 silicon transistor. The operating point is I_C = 10 mA and V_{CE} = 5 V. In addition, V_{CC} = 10 V, β = 330, R_E = 200 Ω, and assume I_{CBO} = 0. (R11-3)

EMITTER-BIAS CIRCUIT WITH TWO SUPPLIES

The emitter-bias circuit with two supplies is shown in figure 11-8. With respect to the dc voltages, this circuit acts more like a common-base circuit than a common-emitter circuit. In general, the voltage drop across R_B is very small because the base current is small. As a result, the base of the transistor is near the ground potential and the emitter lead is below the ground potential. When an ac signal is applied, however, it is coupled to the base of the transistor. Therefore, with respect to the ac voltages, the transistor is in a common-emitter configuration.

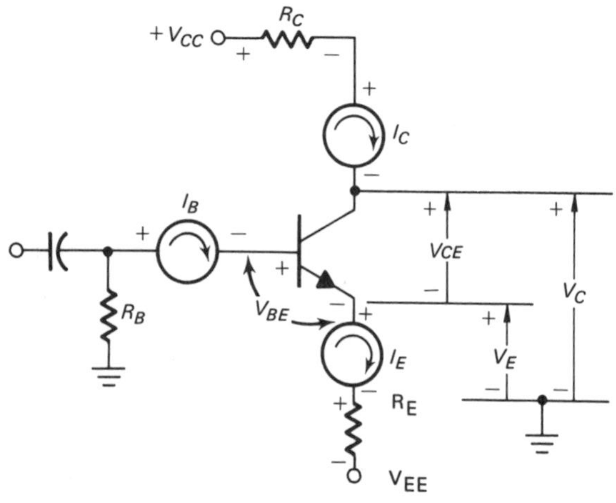

FIG. 11-8 EMITTER-BIAS CIRCUIT WITH TWO SUPPLIES

The circuit in figure 11-8 is very stable since it acts as a common-base circuit with respect to dc or the bias voltages. The collector circuit is practically independent of the transistor β. The collector current is determined by the emitter current. The emitter current is determined almost completely by the emitter supply, V_{EE}, and the emitter resistor, R_E. The emitter supply and the emitter resistor approximate a constant current source because neither component changes easily.

The disadvantage of this circuit is that it requires two power supplies. However, this circuit is very important. If two transistors are used in this configuration with the emitters connected together to one emitter resistor, the result is a differential amplifier commonly used in integrated circuits.

To analyze the circuit in figure 11-8, write the Kirchhoff's voltage expression around the ground-base, resistor-emitter supply circuit.

$$V_{EE} = I_B R_B + V_{BE} + I_E R_E \qquad \text{Eq. 11.20}$$

In the active region, assume that I_{CEO} = 0 and substitute Eq. 7.7 into Eq. 11.20. Solve for the base current I_B as follows:

$$V_{EE} = I_B R_B + V_{BE} + I_B (\beta + 1) R_E$$

$$I_B = \frac{(V_{EE} - V_{BE})}{R_B + (\beta + 1) R_E} \qquad \text{Eq. 11.21}$$

Equation 11.21 can be solved for the collector current by substituting Eq. 7.7 for the base current expressed in terms of the collector current as follows:

$$I_C = \frac{(V_{EE} - V_{BE})}{\frac{R_B}{\beta} + \frac{(\beta+1)}{\beta} R_E} \qquad \text{Eq. 11.22}$$

To determine the transistor voltages, write the Kirchhoff's voltage expression around the collector, supply-collector, resistor-emitter, resistor-emitter supply circuit:

$$V_C = V_{CC} - I_C R_C \quad \text{Eq. 11.23}$$

$$V_{CE} = V_C - V_E \quad \text{Eq. 11.24}$$

The emitter to ground voltage is expressed as:

$$V_E = V_{EE} + I_E R_E$$

$$V_E = V_{EE} + (\beta + 1) I_B R_E \quad \text{Eq. 11.25}$$

Refer to figure 11-8 and note that V_E also can be written as:

$$V_E = -(V_{BE} + I_B R_B) \quad \text{Eq. 11.26}$$

PROBLEM 5.

The circuit in figure 11-8 uses an NPN silicon transistor with $\beta = 100$ and negligible I_{CBO}. Using a 20-V collector supply and a 10-V emitter supply, design an emitter-bias circuit for an operating point of $I_C = 1.0$ mA and $V_{CE} = 10$ V.

For a silicon transistor, $V_{BE} = 0.6$ V.

The circuit in figure 11-8 is designed so that the emitter voltage is near the ground potential. The emitter voltage is expressed as:

$$V_E = I_B R_B + V_{BE}$$

Assuming that $I_B R_B$ is one-tenth as large as V_{BE} (this means that V_{BE} is much larger than $I_B R_B$), V_E becomes:

$$V_E = V_{BE} = 0.6 \text{ V}$$

and $I_B R_B = 0.06$ V

$$\text{Therefore, } R_B = \frac{0.06 \text{ V}}{I_B} = \frac{0.06 \text{ V}}{\frac{I_C}{\beta}} = \frac{0.06 \text{ V}}{\frac{1.0 \text{ mA}}{100}}$$

$$= \mathbf{6.0\ k\Omega}$$

The value of the emitter resistor is determined as follows:

$$R_E = \frac{V_{EE} - V_E}{I_E} = \frac{10 \text{ V} - 0.06 \text{V}}{1.0 \text{ mA}} = \frac{9.4 \text{ V}}{1.0 \text{ mA}}$$

$$= \mathbf{9.4\ k\Omega}$$

The value of the collector resistor is:

$$R_C = \frac{V_{CC} + V_E - V_{CE}}{I_C} = \frac{20\text{V} + 0.6 \text{ V} - 10 \text{ V}}{1.0 \text{ mA}}$$

$$= \frac{10.6 \text{ V}}{1.0 \text{ mA}} = \mathbf{10.6\ k\Omega}$$

It was stated previously in this section that this circuit is nearly independent of the transistor β. To test this statement, increase β to five times the given value in Problem 5 and calculate I_C using Eq. 11.22.

$$I_C = \frac{(V_{EE} - V_{BE})}{\frac{R_B}{\beta} + \frac{(\beta + 1)}{\beta} R_E}$$

$$= \frac{10 - 0.6}{\frac{6000}{500} + \frac{501}{500}(9400)} = \frac{9.4}{12 + 9400}$$

$$= \frac{9.4 \text{ V}}{9412\ \Omega} = 1.0 \text{ mA}$$

It can be seen that I_C does not change value for an increase (or decrease) in the transistor β for the emitter-bias circuit with two supplies.

The actual value for R_B has little effect on the collector current. Referring back to Problem 5, if the value for R_B is increased ten times to 60 kΩ, the collector current can be calculated using Eq. 11.22.

$$I_C = \frac{10 - 0.6}{\frac{60000}{100} + \frac{101}{100}(9400)} = 0.94 \text{mA}$$

There is less of a change in the collector current for this circuit than is found in the other biasing circuits covered. For example, if R_B is increased ten times in the fixed-bias circuit, I_C decreases by one-tenth. For the circuit in Problem 5, when R_B is increased ten times, the collector current decreases only six percent of its original value.

LABORATORY EXERCISE 11-1: EMITTER-BIAS CIRCUIT WITH VOLTAGE DIVIDER

PURPOSE

- To design an emitter-bias CE circuit with voltage divider for a specified operating point.
- To construct an emitter-bias circuit in the laboratory and measure the β of the transistor and the circuit operating point.
- To compare actual laboratory data with the theoretical calculations.

MATERIALS

1 NPN transistor, 2N4425, 2N3405 or equivalent
1 Dc power supply, 10 V
1 VTVM, or solid-state voltmeter
2 VOMs, 20,000 ohms per volt
1 Resistor, 510 Ω, ½W
1 Resistor, 300 k Ω, ½W
Miscellaneous ½-W resistors to be determined in step A of the Procedure

PROCEDURE

A. 1. Draw an emitter-bias CE circuit with a voltage divider using an NPN transistor. Label all components. Insert ammeters to measure I_B and I_C. Show the voltages V_C, V_E, V_B (the voltage from base to ground), V_{CE}, and V_{BE}. Using a 10-V supply with R_E = 200 Ω and R_B = 1800 Ω, perform the necessary calculations to establish an operating point of I_C = 10 mA and V_{CE} = 5 V. Assume that V_{BE} = 0.6 V and I_{CBO} is negligible.

2. Determine R_C. Record this value in Table 11-2, step A.2.

3. Determine and record the values of I_B and V_{OC} for each value of β in Table 11-2, steps A.2. and A.3.
$V_{OC} = I_B\,[R_B + (\beta + 1)\,R_E] + V_{BE}$

TABLE 11-2

Procedure Step	Comments	β	R_C in Ω	R_E in Ω	R_1 in Ω	R_2 in Ω	I_C in mA	V_{OC} in V	V_C in V	V_E in V	V_{CE} in V	I_B in μA	I_E in mA	V_B in V	V_{BE} in V
A.2	Theoretical Calculations	330		200			10		7	2	5		–	2.6	0.6
A.3	Preliminary Work	250													
A.3	Preliminary Work	150													
B.1	Actual β			–	R_B =				–	–			–	–	
C.2	Actual Lab Data	–						–				–			

4. Note in step A.3. that β has little effect on V_{OC}. The emitter-bias CE circuit with voltage divider is beta independent. Use the value of V_{OC} for β = 330 to calculate the values of R_1 and R_2 in the voltage divider circuit. Record these values in Table 11-2, step A.2.

$$R_1 = \frac{R_B\ V_{CC}}{V_{OC}} \qquad R_2 = \frac{R_B\ V_{CC}}{V_{CC} - V_{OC}}$$

B. 1. Connect the fixed-bias CE circuit shown in figure 11-9.

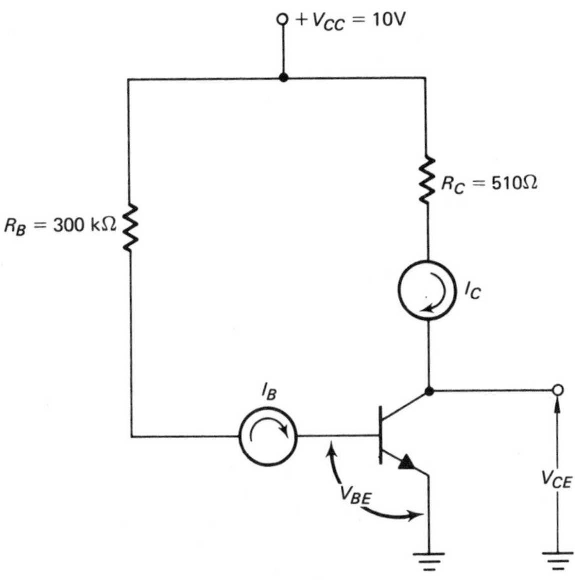

FIG. 11-9 FIXED-BIAS CIRCUIT TO DETERMINE β OF THE TRANSISTOR

2. Determine and record all of the data required for step B.1. of Table 11-2. Determine the actual β of the transistor using the measured values for I_C and I_B.

C. 1. Connect the circuit drawn in step A.1.

2. Use the values of V_{CC}, R_C, R_E, R_1, and R_2 as given or determined in step A. Use resistor values as close to calculated values as possible. Two resistors connected in series or in parallel may be used if necessary. *Do not use potentiometers.* Record the values of the resistors used in step C.2. of Table 11-2.

   CAUTION: Do not insert an ammeter to measure I_B. Most ammeters have a very high resistance for the μA scales and this high resistance will change the biasing circuit.

D. Determine and record all of the remaining data required for step C.2. of Table 11-2.

DISCUSSION QUESTIONS

1. Compare the theoretical and actual test results of steps A.2. and C.2. in Table 11-2. Discuss the results and give reasons for any discrepancies.
2. Using the measured values of step C.2., verify Kirchhoff's voltage law around the base-emitter-ground circuit loop and the collector-emitter-ground circuit loop.
3. Does the operating point of the emitter-bias with voltage divider CE amplifier circuit depend upon the β of the transistor? Discuss and use laboratory data to substantiate this answer.
4. Step C of the Procedure cautioned against using an ammeter to read I_B. Determine the resistance of a VOM on the μA scale required for the measurement of I_B. Draw the emitter-bias with voltage divider CE amplifier circuit and insert the VOM resistance for the μA range in place of the I_B ammeter. Using this circuit, determine the effect of an I_B ammeter on the biasing circuit.
5. If the fixed-bias and emitter-bias CE amplifier circuits are used in the laboratory exercises of unit 11, compare the results obtained using these circuits with the results obtained from the emitter-bias with voltage divider CE amplifier circuit. For which circuit is the actual operating point closest to its theoretical operating point? Why? Which circuit is *less* β dependent? Why?

LABORATORY EXERCISE 11-2: COLLECTOR-BASE BIAS CIRCUIT

PURPOSE

- To design a collector-base bias CE circuit with and without an emitter resistor for a specified operating point.
- To construct collector-base bias circuits in the laboratory and measure the operating point of the circuits and the β of the transistor.
- To observe the variation of the operating point with changes in the base resistor.
- To compare actual laboratory data with the theoretical calculations.

MATERIALS

1 NPN transistor, 2N4425, 2N3405 or equivalent
1 Dc power supply, 10 V
1 VTVM, or solid-state voltmeter
2 VOMs, 20,000 ohms per volt
Miscellaneous ½-W resistors to be determined in step A of the Procedure

PROCEDURE

A. 1. Draw the collector-base bias circuits (with and without emitter resistors) using an NPN transistor. Label all components. Insert ammeters to measure I_B and I_C. Show the voltages V_C, V_E, V_{CE}, and V_{BE}. Using a 10-V supply and with R_E = 200 ohms when used, perform the following calculations to establish an operating point of I_C = 10 mA and V_{CE} = 5 V. Assume that V_{BE} = 0.6 V and I_{CBO} is negligible.

TABLE 11-3

β	I_B in μA	Collector-Base Bias	
		Without R_E R_B in Ω	With R_E R_B in Ω
330			
250			
150			

2. Determine R_C for both circuits. Record these values in the theoretical section of Table 11-4.

TABLE 11-4

Step	Comments	β	R_C in Ω	R_E in Ω	R_B in Ω	I_C in mA	V_C in V	V_E in V	V_{CE} in V	I_B in μA	V_{BE} in V
—	Theoretical without R_E	330		—		10	—	—	5		0.6
—	Theoretical with R_E	330		200		10	7	2	5		0.6
C	Lab Data without R_E			—			—	—			
D	Vary R_B			—			—	—			
F	Lab Data with R_E										
G	Vary R_B										

3. Determine and record the values of I_B and R_B for each value of β given in Table 11-3.

B. Connect the circuit drawn in step A.1. but omit the emitter resistor. Use the values of V_{CC}, R_C, and R_B (for β = 330) as determined or given in step A. Use resistor values as close to the calculated values as possible. Two resistors connected in series or in parallel may be used if necessary. *Do not use potentiometers.* Record the values of the resistors used in step C. of Table 11-4.

C. Complete all of the data required for step C. Table 11-4. Determine the actual value of β for the transistor by using the measured values for I_C and I_B.

D. If the V_{CE} of the operating point obtained in step C. does not have a value between 4 V and 6 V, change the base resistor so that V_{CE} is approximately 5 V. Complete step D. in Table 11-4.

E. If step D. is performed, disregard step E. and go to step F. If the V_{CE} of the operating point is between 4 V and 6 V, change R_B to the value found in step A.3. for $\beta = 150$ to observe the effect on the operating point of changing resistance. Complete step D of Table 11-4.

F. Connect the circuit drawn in step A.1., including the emitter resistor. Use the values of V_{CC}, R_C, R_E, and R_B for $\beta = 330$ as determined or given in step A. Use resistor values as close to the calculated values as possible. Two resistors connected in series or in parallel may be used if necessary. *Do not use potentiometers.* Record the values of the resistors used in step F. of Table 11-4.

G. Repeat steps C, D, and E for the collector-base bias circuit with an emitter resistor. Record all data in step G. of Table 11-4.

DISCUSSION QUESTIONS

1. Compare the theoretical and actual test results listed in Table 11-4. Discuss the results and give reasons for any discrepancies.
2. Using the measured values of step C, verify Kirchhoff's Voltage Law around the base-emitter circuit loop and the collector-emitter circuit loop. Repeat this procedure using the measured values of step F.
3. Does the operating point in the collector-base bias CE amplifier circuits depend upon the β of the transistor? Discuss and use laboratory data to substantiate this anwer.
4. Compare the results obtained for the circuits with and without the emitter resistor. For which circuit is the actual operating point closer to its theoretical operating point? Why? Which circuit is *less* β dependent? Why?

LABORATORY EXERCISE 11-3: EMITTER-BIAS CIRCUIT WITH TWO SUPPLIES

PURPOSE

- To design an emitter-bias CE circuit with two supplies for a specified operating point.
- To construct an emitter-bias circuit with two supplies and measure the β of the transistor and the circuit operating point.
- To observe the variation of the operating point with changes of the base resistor.
- To compare the actual laboratory data with the theoretical calculations.

MATERIALS

1 NPN transistor, 2N4425, 2N3405 or equivalent
1 Dc power supply, 10 V
1 Dc power supply, 5 V
1 VTVM, or solid-state voltmeter
2 VOMs, 20,000 ohms per volt
Miscellaneous ½-W resistors to be determined in the Procedure

PROCEDURE

A. 1. Draw an emitter-bias CE circuit with two supplies using an NPN transistor. Label all components. Insert ammeters to measure I_B and I_C. Show voltages, V_C, V_E, V_B, (voltage from base to ground), V_{CE}, and V_{BE}. Using a 10-V supply for V_{CC} and a 5-V supply for V_{EE}, perform the following calculations to establish an operating point of $I_C = 10$ mA and $V_{CE} = 5$ V. Assume that $V_{BE} = 0.6$ V and I_{CBO} is negligible.

2. Determine R_C and R_E. Record these values in Table 11-5, step A.2.

TABLE 11-5

Procedure Step	Comments	β	R_C in Ω	R_E in Ω	R_B in Ω	I_C in mA	V_C in V	V_E in V	V_{CE} in V	I_B in μA	V_B in V	V_{BE} in V
A.2	Theoretical Calculations for Preliminary Work	330				10	4.4	-0.6	5		0.6	0.6
A.3		250										
A.3		150										
A.4	R_B ten times larger	330										
B.2	Actual Lab Data											
D	Vary R_B (β= 150)											
E	R_B ten times bigger											

3. Determine the values of I_B and R_B for each value of β in Table 11-5 for steps A.2. and A.3. Let the product $I_B R_B = 0.06$ V. Record the values of I_B and R_B for $\beta = 330$ in the theoretical calculation section of Table 11-6 marked step A.2.

4. Increase the value of the base resistor R_B to ten times the determined value for step A.2. Using this value for R_B, determine all of the values required for step A.4. of Table 11-5.

B. 1. Connect the circuit drawn in step A.1. Use the values of V_{CC}, V_{EE}, R_C, and R_E as determined or given in step A. Assume β = 330 and use the value determined for R_B in step A.2. Use resistor values as close to the calculated values as possible. Two resistors connected in series or in parallel may be used if necessary. *Do not use potentiometers.*

2. Record the values of the resistors used in Table 11-5, step B.2.

C. Complete and record all of the data required for step B.2. of Table 11-5. Determine the actual value of β for the transistor by using the measured values of I_C and I_B.

D. Change R_B to the value found in step A.3. for $\beta = 150$ to observe the effect of changing resistance on the operating point. Complete step D. in Table 11-5.

E. Change R_B to the value found in step A.4. Complete step E in Table 11-5.

DISCUSSION QUESTIONS

1. Compare the theoretical and actual test results of steps A.2 and B.2 in Table 11-5. Discuss the results and give reasons for any discrepancies.
2. Using the measured values of step B.2., verify Kirchhoff's voltage law around the base-emitter-ground circuit loop and the power supplies-collector-emitter-ground circuit loop.
3. Does the operating point in the emitter-bias CE amplifier circuit with two supplies depend upon the β of the transistor? Discuss and use laboratory data to substantiate this answer.
4. Does the operating point in the emitter-bias CE amplifier circuit with two supplies depend upon the base resistor R_B? Discuss and use laboratory data to substantiate this answer.
5. What is the resistance of the laboratory VOM on the μA scale used in the measurement of I_B? Does this resistance affect the biasing circuit of the transistor?
6. List the operating points found for each of the six biasing circuits in Table 11-6. Compare and discuss the results.

TABLE 11-6

Circuit	Operating Point	
	V_{CE}	I_C
Fixed-bias		
Collector-base bias		
Collector-base bias with emitter resistor		
Emitter-bias		
Emitter-bias with voltage divider		
Emitter-bias with two supplies		

EXTENDED STUDY TOPICS

1. For the emitter-bias circuit with voltage divider in figure 11-5, find the missing quantities in Table 11-7. Neglect I_{CBO}.

TABLE 11-7

	Type	β	V_{BE} in V	V_{CC} in V	R_1 in kΩ	R_2 in kΩ	R_C in k Ω	R_E in k Ω	I_C in mA	I_E in mA	I_B in μA	V_{CE} in V	V_{CG} in V
a	NPN	40	0.6	+20			8.0	2.0			25		
b	PNP	90	0.2	-6	10	4.7	0.4	0.22					
c	NPN	500	0.6	+20	76	12.0	4.0				4	10	
d	NPN	330	0.6	+10	76	12.0		0.2	10			5	

2. Design a collector-base bias common emitter transistor using a 2N4425 silicon transistor. The operating point is I_C = 10 mA and V_{CE} = 5 V. In addition, V_{CC} = 10 V, β = 330, and I_{CBO} = 0.
3. Determine the operating point for the circuit in figure 11-6 if R_B = 220 kΩ, R_C = 5 kΩ, V_{CC} = 10 V, β = 50, V_{BE} = 0.6 V, V_{CE} = 5 V, and I_{CBO} = 0.
4. Determine the operating point for the circuit in figure 11-6 if R_B = 100 kΩ, R_C = 10 kΩ, V_{CC} = 18 V, β = 100, V_{BE} = 0.2 V, V_{CE} = 9 V and I_{CBO} = 0.
5. Determine the operating point for the circuit in figure 11-7 if R_B = 220 k Ω, R_C = 4 k Ω, R_E = 1 k Ω, V_{CC} = 10 V, β = 50, V_{BE} = 0.6 V, V_{CE} = 5 V, and I_{CBO} = 0.
6. Determine the operating point for the circuit in figure 11-7, if R_B = 420 kΩ, R_C = 5.1 kΩ, R_E = 1 kΩ, β = 50, V_{BE} = 0.6 V, V_{CE} = 10 V, V_{CC} = 20 V, and I_{CBO} = 0.
7. Design a two-supply emitter-bias transistor circuit using a 2N4425 silicon transistor. The operating point is I_C = 10 mA and V_{CE} = 5 V. In addition, V_{CC} = 10 V, V_{EE} = 5 V, β = 330, R_E = 220 Ω and I_{CBO} = 0.
8. Determine the operating point for the circuit in figure 11-8, if R_B = 3 k Ω, R_C = 5.6 k Ω, R_E = 4.4 k Ω, V_{CC} = 10 V, V_{EE} = 5 V, β = 50, V_{BE} = 0.6 V, and I_{CBO} = 0.

Appendix

ANSWERS TO STUDENT REVIEW QUESTIONS

Unit 1 Introduction to *p-n* Junctions and the Semiconductor Diode

(R1-1) PE = -2.56×10^{-19} J; KE = $+1.28 \times 10^{-19}$ J

(R1-2) v = 5.3×10^{5} m/s

(R1-3)
a. The bonding of atoms by valence electrons transferring from one atom to another
b. The bonding of atoms when sharing of valence electrons takes place
c. The bonding of atoms when valence electrons float in a cloud among positive ions of metallic atoms

(R1-4) Germanium is more sensitive to heat since its forbidden gap is smaller than that of silicon. Electrons from the valence band of germanium require less energy to move from the valence band to the conduction band

(R1-5) σ = 0.0232 mhos/cm
ρ = 43.1 ohm-cm

(R1-6) $\sigma = 5.12 \times 10^{-6}$ mhos/cm
$\rho = 195.3 \times 10^{3}$ ohm-cm

(R1-7)
a. Controlled addition of impurity atoms
b. Pure germanium
c. Electrons
d. Holes
e. Group V elements

(R1-8) A group V element forms a covalent bond with germanium neighbors; however, since the group VI element has five valence electrons, the fifth valence electron doesn't participate in the covalent bond

(R1-9)
a. Holes
b. Electrons
c. Movement of holes through a solid structure
d. Group III elements

(R1-10) A group III element forms a covalent bond with germanium neighbors; however, since the group III element has three valence electrons, a hole is created in the bonding process

(R1-11) Grown junction, alloy junction, epitaxial layer, diffusion, and electrochemical etching

(R1-12) Ions are formed about the junction by electrons and holes combining around the junction. The ions create a barrier of depleted charge carriers which prohibits further electron-hole combinations

(R1-13)
a. Group V atoms which have lost an electron
b. The depletion region around the junction
c. The area around the junction where charge carriers (electrons and holes) have been removed

(R1-14) The forward-biased junction has a plus voltage on the p side and a minus voltage on the n side. This voltage polarity effectively makes the depletion region become smaller and causes charge carriers to cross the junction and current to flow. The reverse-biased junction has a minus voltage on the p side and a plus voltage on the n side. The voltage polarity effectively makes the depletion region larger and makes it harder for charge carriers to cross the junction. Therefore, there is no current flow.

(R1-15) Minority current flow

(R1-16) Silicon has a larger forbidden gap than germanium

Unit 2 Rectification

(R2-1) a. V_{dc} = 127.2 V b. V_{PRV} = 400 V c. I_{dc} = 57.8 mA
(R2-2) V_{PRV} = 314 V
(R2-3) a. V_{dc} = 56 V b. V_{PRV} = 176.1 V
(R2-4) V_s = 2400 V I_p = 24A
I_s = 1.2 A R_p = 5Ω
(R2-5) V = 3.33 V I_p = 0.555 mA
I_s = 6.66 mA R_p = 72 kΩ
(R2-6) V_s = 3600 V I_p = 54 A
I_s = 1.8 A R_p = 2.22 Ω
(R2-7) V_{dc} = 224.8 V I_{dc} = 224.8 mA V_{PRV} = 707 V
(R2-8) V_{dc} = 2158 V I_p = 157 mA R_p = 54 Ω V_{PRV} = 6787.2 V

(R2-9)

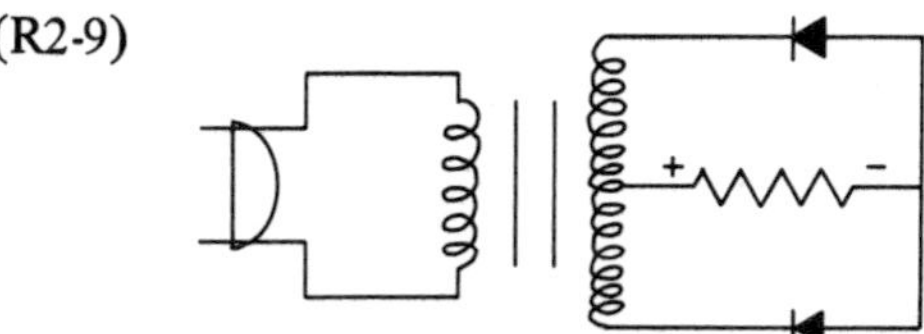

(R2-10) V_{PRV} = 118 V
(R2-11) V_s = 1200 V $V_{dc\,(fw)}$ = 1079 V V_{PRV} = 1696.8 V I_{dc} = 169.68 mA

Unit 3 Power Supply Filters

(R3-1) $\%\gamma$ = 0.3535%
(R3-2) Measure dc load voltage and rms load voltage

(R3-3) R_L = 40 kΩ; $\% \frac{R_s}{R_L}$ = 2.5%; ωCR_L = 150; V_p = 116.3 V; V_{rms} = 82.2 V;

PRV = 116.3 V; I_s = 116.3 mA; I_p = 19.75 mA

(R3-4) R_L = 12.5 kΩ; $\% \frac{R_s}{R_L}$ = 8%; ωCR_L = 0.471; V_p = 169.5 V, V_{rms} = 120 V;

PRV = 239 V; I_s = 169.5 mA; I_p = 13.6 mA

(R3-5) R = 25 kΩ; $\% \frac{R_s}{R_L}$ = 4%; ωCR_L = 0.942; V_p = 312.5 V, V_{rms} = 221 V;

PRV = 312.5 V; I_s = 312.5 mA; I_p = 13.6 mA

(R3-6) R_L = 13.3 kΩ; $\% \frac{R_s}{R_L}$ = 7.5%; ωCR_L = 25; V_p = 84.75 V, V_{rms} = 60 V;

PRV = 120 V; I_p = 75 mA

(R3-7) R_L = 80 kΩ; $\% \frac{R_s}{R_L}$ = 1.25%; ωCR_L = 15; V_p = 69.4 V, V_{rms} = 49 V;

PRV = 98 V; I_p = 13 mA
(R3-8) V'_{dc} = 56.1 V
(R3-9) V'_{dc} = 119.7 V
(R3-10) R_L = 2 kΩ; L_C = 2.123 H; yes
(R3-11) f_{in} = 42.5 Hz for an L_C = 2.5 H Any frequency below 42.5 Hz makes L = 2.5 H not a critical inductance

Unit 4 Zener Diodes

(R4-1) Junction breakdown and avalanche breakdown
(R4-2) Junction phenomenon at lower voltages; avalanche phenomenon at high voltages
(R4-3) Reference diode possesses only the Zener voltage characteristics
(R4-4) $I_{R_1} = 50$ mA; $I_{R_2} = 450$ mA; $I_{ZT} = 250$ mA
(R4-5) $Z_1 = 360\ \Omega$; $Z_2 = 48.89\ \Omega$
(R4-6) $V^1{}_{ZT} = 15.9$ V
(R4-7) $V^1{}_{ZT} = 5.6882$ V
(R4-8)
a. Use Zener diode 1N1766
b. $I_{ZM} = 161.3$ mA; $I_{R_1} = 1613$ mA
c. $I_{Z\,(MID)} = 88.715$ mA
d. $I_T = 138.715$ mA
e. $V_s = 16.86$ V
f. $R_s = 121.5\ \Omega$; $P_s = 2.34$ W

Unit 5 Junction Transistor Familiarization

(R5-1)

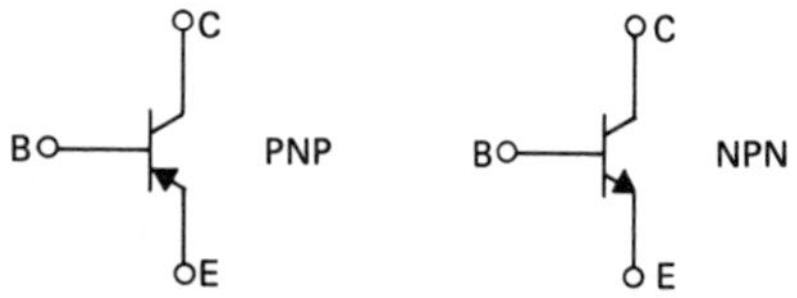

C
B PNP
E

C
B NPN
E

(R5-2) The emitter, base, and collector are the parts of a transistor. The emitter section provides charge carriers; the base section controls the flow of charge carriers; and the collector collects charge carriers.

(R5-3) Holes are to be moved from the emitter to the base. To do this, the emitter must be at (+) voltage and the base at (-) voltage, and the collector (-) voltage with respect to the base (+) voltage. The emitter holes will move from the emitter region to the base region where approximately 5% of the holes will stay in the base and about 95% of the holes will move to the collector

(R5-4) $I_E = I_B + I_C$

Unit 6 Common-Base Configuration

(R6-1) $\gamma = -0.943$; $I_C = 33$ mA; $I_B = 2$ mA
(R6-2) $I_E = 1.714$ mA; $I_C = 1.604$ mA
(R6-3) $\gamma_{ac} = -0.94$
(R6-4) $\triangle I_E = 0.333$ mA; $\triangle I_C = 0.326$ mA
(R6-5) $A_V = 1500$

Unit 7 Common-Emitter and Common-Collector Amplifiers

(R7-1) $\beta = 162.5$, $I_E = 65.4$ mA; $\beta = 100$, $I_E = 101$ mA; $\beta = 175$, $I_E = 140.8$ mA; $\beta = 212$, $I_E = 85.4$ mA
(R7-2) $I_{CS} = 4.55$ mA, $I_B = 0.0455$ mA
(R7-3) $V_{CE} = 22$ V; $V_B = 6.4$ V; $I_B = 0.044$ mA; $I_C = 1.956$ mA

Unit 8 Transistor Specifications and Graphical Analysis

(R8-1) $P_C\,(40°C) = 100.05$ mW; $P_C\,(55°C) = 50.1$ mW; $P_C\,(100°C) = -99.75$ mW
(R8-2) $P_{max} = V_{CE}\,I_C$. The maximum power dissipation curve indicates the units of the current for a particular voltage and vice versa.

Unit 9 Transistor Load Lines

(R9-1) V_{CC}, the collector supply voltage source

(R9-2) $I_C = \frac{V_{CC}}{R_C}$

(R9-3) I_B, I_C, and V_{CE}

(R9-4) $V_{CE} = V_{CC} = 20$ V, $I_C = 0$
$V_{CE} = 0$, $I_C = 10$ mA

(R9-5) The Q point is common to both the dc and ac load lines

Unit 11 More Common-Emitter Biasing

(R11-1) $I_C = 1$ mA; $I_E = 1.02$ mA; $V_{CE} = 4.98$ V; $V_{CG} = 6$ V

(R11-2) $I_B = 15.2\ \mu A$; $R_B = 0.224\ M\Omega$; $R_C = 1.2\ k\Omega$

(R11-3) $I_B = 30\ \mu A$; $R_C = 0.3\ k\Omega$; $R_B = 0.313\ M\Omega$

Acknowledgments

Source Editor: Marjorie A. Bruce, Technical Education Division

Technical Review provided by: Charles Peterpaul, Hudson Valley Community College, Troy, New York

Consulting Editor, Electronic Technology Series: Richard L. Castellucis

The staff at Delmar Publishers:

Director of Publications: Alan N. Knofla

Director of Manufacturing and Production: Frederick Sharer

Production Specialists: Sharon Lynch, Jean LeMorta, Patti Manuli, Betty Michelfelder, Debbie Monty, Lee St. Onge, Alice Schielke

Illustrators: Anthony Canabush, George Dowse, Michael Kokernak

Appreciation is expressed to the following companies and organizations for their permission to use the technical data noted.

General Electric Company, Semiconductor Products Department: Figures 4-5, 8-1, 8-2

Institute of Electrical and Electronics Engineers, Inc.: Figures 3-6, 3-7, 3-8, 3-9, 5-8, 6-3

Index

A

AC load line, 155-56
Active region, 84, 109
Amplifier input resistance, 97, 114
Amplifier output resistance, 98, 114
Avalanche theory, 62

B

Bands of energies, 6
Barrier, 14
Barrier width, 14
Bonding, 4
Breakdown diode, 64
Breakdown voltage, 62

C

Characteristic curve, 15
Choke, 52
Choke input filter, 52-53
Collector-base bias circuit, 187
Collector-base junction, 77
Collector characteristics, 106-108
Collector dissipation, 143
Common-base amplifiers, 93-95
Common-base characteristic curves, 82-83
Common-base circuit, 81
Common-base current gain, 81
Common-collector amplifiers, 115-118
Common collector circuit, 81
Common-emitter amplifiers, 109-112
Common-emitter characteristics curves, 105-109
Common emitter circuit, 81
Conduction band, 6-8
Coulomb's law for charged bodies, 1
Covalent bonding, 4
Critical inductance, 55
Cutoff region, 84, 109

D

DC load line, 150-151
Depletion region, 14
Derating factors, 143
Diode, 21-24
Doped with holes, 12

E

Electron forces, 1-3
Electron-hole pairs, 10
Electron-pair bonding, 5
Electrons, 8
Electrostatic potential energy, 3
Emitter, 105
Emitter-base junction, 77
Emitter-bias circuit
  with single base resistor, 174-175
  with two supplies, 188-189
  with voltage divider, 181-183
Emitter follower, 115-116
Energy bands, 6-8
Energy level diagram, 6

F

Fixed-bias circuit, 171-172
Forbidden band, 6-8
Forward bias, 15
Fourier analysis, 54
Free electron, 11
Full-wave bridge rectifier circuit, 30-32

H

Holes, 8

I

Ideal transformer theory, 25-27
Input characteristic, 82, 105
Input resistance, 97
Ionic bonding, 4

J

Junction breakdown, 62
Junction transistor action, 77-81

K

KE *See* Kinetic energy
Kinetic energy, 2

L

Leakage current, 15, 84, 109
L filters, 55
L section filter, 52-53

M

Majority carriers, 12

Maximum allowable dc output current, 22
Metallic bonding, 4
Minority carriers, 12, 78
Mobility factor, 8
Molecular bonding, 4-5

N

Nonlinear amplifier circuit, 168
N type semiconductor, 11-12

O

Operating current, 69
Operating point, 150
Operating temperature, 143
Output characteristic, 82, 105, 106-108
Output resistance, 97

P

PE *See* Electrostatic potential energy
Pentavalent elements, 11
Percent ripple factor, 44-45
Photons, 6
P-n junction, 13-16
Pn junction diode, 13

S

Secondary turns, 25
Semiconductor, 8
Semiconductor diode, 16-17
Solid-state switch, 171
Super-beta transistors, 109

T

Thermal resistance, 143-144
Thevenin's Equivalent circuit, 181
Total transistor dissipation, 143
Transistor action, 78
Transistor amplifier, 168
Transistor load lines, 149-152
Transistor specification sheets, 132-144
Trivalent elements, 11
Two-diode full-wave rectifier, 27-29

V

Vacancy, 8
Valence, 4
Valence band, 6-8
Voltage doubler circuit, 50
Voltage regulation, 46

Z

Zener breakdown, 62
Zener diodes, 62-69
Zener diode voltage regulator circuits, 70-71
Zener voltage, 62

777 (6C1960)